U.S. Marine Corps Close Combat

Plus

Mountain Operations

Department of Defense

Pentagon
Publishing

DEPARTMENT OF THE NAVY
Headquarters United States Marine Corps
Washington, D.C. 20380-1775

18 February 1999

FOREWORD

1. PURPOSE

Today's Marines operate within a continuum of force where conflict may change from low intensity to high intensity over a matter of hours. Marines are also engaged in many military operations other than war, such as peacekeeping missions or noncombatant evacuation operations, where deadly force may not be authorized. During noncombative engagements, Marines must determine if a situation warrants applying deadly force. Sometimes Marines must decide in a matter of seconds because their lives or the lives of others depend on their actions. To make the right decision, Marines must understand both the lethal and nonlethal close combat techniques needed to handle the situation responsibly without escalating the violence unnecessarily. Marine Corps Reference Publication (MCRP) 3-02B, *Close Combat,* provides the tactics, techniques, and procedures of Marine Corps close combat. It also provides the doctrinal basis for the Marine Corps Close Combat Training Program (MCCCTP).

2. SCOPE

This publication guides individual Marines, unit leaders, and close combat instructors in the proper tactics, techniques, and procedures for close combat. MCRP 3-02B is not intended to replace supervision by appropriate unit leaders and close combat instruction by qualified instructors. Its role is to ensure standardization and execution of tactics, techniques, and procedures throughout the Marine Corps.

3. SUPERSESSION

MCRP 3-02B supersedes Fleet Marine Force Manual (FMFM) 0-7, *Close Combat,* dated 9 July 1993. There are significant differences between the two publications. MCRP 3-02B should be reviewed in its entirety.

4. WARNING

Techniques described in this publication can cause serious injury or death. Practical application in the training of these techniques will be conducted in strict accordance with approved Entry Level Close Combat, Close Combat Instructor (CCI), and Close Combat Instructor Trainer (CCIT) lesson plans. Where serious danger exists, the reac er is alerted by the following:

WARNING

---

5. CERTIFICATION

Reviewed and approved this date.

BY DIRECTION OF THE COMMANDANT OF THE MARINE CORPS

J. E. RHODES
Lieutenant General, U.S. Marine Corps
Commanding General
Marine Corps Combat Development Command

DISTRIBUTION: 14400006600

OVERVIEW OF CLOSE COMBAT

1. Purpose of Close Combat

Close combat is the physical confrontation between two or more opponents. It involves armed and unarmed and lethal and nonlethal fighting techniques that range from enforced compliance to deadly force. The purpose of close combat is to execute armed and unarmed techniques to produce both lethal and nonlethal results. Unarmed techniques include hand-to-hand combat and defense against hand-held weapons. Armed techniques include techniques applied with a rifle, bayonet, knife, baton, or any weapon of opportunity.

2. Continuum of Force

Marines will find themselves in both combative and noncombative situations. The threat level in these situations can rise and fall several times based on the actions of both Marines and the people involved. The escalation of force stops when one person complies with the demands imposed by another person. This range of actions is known as a continuum of force. Continuum of force is the concept that there is a wide range of possible actions, ranging from voice commands to application of deadly force, that may be used to gain and maintain control of a potentially dangerous situation (MCO 5500.6_, *Arming of Security and Law Enforcement [EE] Personnel and the Use of Force).* Continuum of force consists of five levels that correspond to the behavior of the people involved and the actions Marines use to handle the situation (see the table below). Close combat techniques are executed in levels three, four, and five.

Level One: Compliant(Cooperative)

The subject complies with verbal commands. Close combat techniques do not apply.

Level Two: Resistant (Passive)

The subject resists verbal commands but complies immediately to any contact controls. Close combat techniques do not apply.

Continuum of Force		
Level	**Description**	**Actions**
1	Compliant (Cooperative)	Verbal commands
2	Resistant (Passive)	Contact controls
3	Resistant (Active)	Compliance techniques
4	Assaultive (Bodily Harm)	Defensive tactics
5	Assaultive (Serious Bodily Harm/ Death)	Deadly force
Note: Shading indicates levels in which Marines use close combat techniques.		

Level Three: Resistant (Active)

The subject initially demonstrates physical resistance. Marines use compliance techniques to control the situation. Level three incorporates close combat techniques to physically force a subject to comply. Techniques include—

- Come-along holds.
- Soft-handed stunning blows.
- Pain compliance through joint manipulation and the use of pressure points.

Level Four: Assaultive (Bodily Harm)

The subject may physically attack Marines, but he does not use a weapon. Marines use defensive tactics to neutralize the threat. Defensive tactics include the following close combat techniques:

- Blocks.
- Strikes.
- Kicks.
- Enhanced pain compliance procedures.
- Nightstick blocks and blows.

Level Five: Assaultive (Serious Bodily Harm/Death)

The subject usually has a weapon and will either kill or seriously injure someone if he is not stopped immediately and brought under control. Typically, to control the subject, Marines apply deadly force through the use of a firearm, but they may also use armed and unarmed close combat techniques.

3. Marine Corps Tactical Concepts

Close combat techniques support the following key Marine Corps tactical concepts. The concepts are not standalone ideas but are to be combined to achieve an effect that is greater than their separate sum.

Achieving a Decision

Achieving a decision is important in close combat. An indecisive fight wastes energy and possibly Marines' lives. Whether the intent is to control an opponent through restraint or defend themselves in war, Marines must have a clear purpose before engaging in close combat and act decisively once engaged.

Gaining an Advantage

A basic principle of martial arts is to use the opponent' s strength and momentum against him to gain more leverage than one's own muscles alone can generate, thereby gaining an advantage. In close combat, Marines must exploit every advantage over an opponent to ensure a successful outcome. This can include employing various weapons and close combat techniques that will present a dilemma to an opponent. Achieving surprise can also greatly increase leverage. Marines try to achieve surprise through deception, stealth, and ambiguity.

Speed

Marines use speed to gain the initiative and advantage over the enemy. In close combat, the speed and violence of the attack against an opponent provides Marines with a distinct advantage. Marines must know and understand the basics of close combat so they can act instinctively with speed to execute close combat techniques.

Adapting

Close combat can be characterized by friction, uncertainty, disorder, and rapid change. Each situation is a unique combination of shifting factors that cannot be controlled with precision or certainty. For example, a crowd control mission may call for Marines to employ various techniques ranging from nonlethal restraint to more forceful applications. Marines who adapt quickly will have a significant advantage.

Exploiting Success

Typically, an enemy will not normally surrender simply because he was placed at a disadvantage. Marines cannot be satisfied with gaining an advantage in a close combat situation. They must exploit any advantage aggressively and ruthlessly until an opportunity arises to completely dominate the opponent. Marines must exploit success by using every advantage that can be gained.

(reverse blank)

Close Combat

Table of Contents

Page

CHAPTER 1

FUNDAMENTALS OF CLOSE COMBAT

This chapter describes all techniques for a right-handed person. However, all techniques can be executed from eittwr side.

The Marine is depicted in camouflage utilities. The opponent is depicted without camouflage.

The fundamentals of close combat include ranges, weapons of the body, target areas of the body, and pressure points of the body. These fundamentals form the basis for all close combat techniques. They provide Marines with a common framework regardless of the type of confrontation or the techniques used. If Marines apply these fundamentals properly in a close combat situation, they may save their lives or the lives of fellow Marines.

1. Ranges of Close Combat

Close combat engagements occur within three ranges: long range, midrange, and close range. During any engagement, these ranges may blur together or may rapidly transition from one to another until either the opponent is defeated or the conflict is resolved.

Long Range

During long range engagements, combatants engage each other with rifles, bayonets, sticks, or entrenching tools. See figure below.

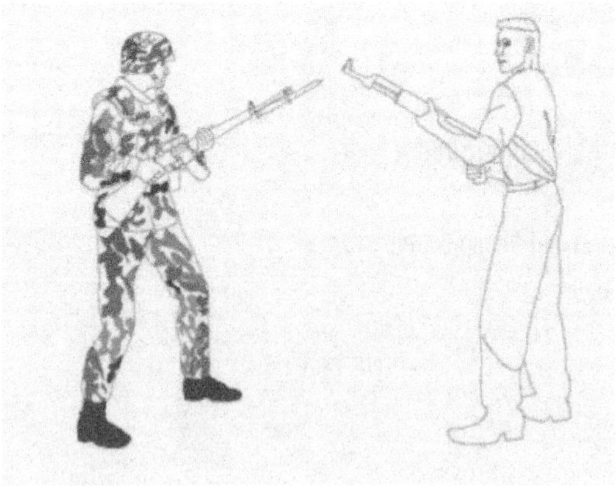

Midrange

During midrange engagements, combatants engage each other with knives, punches, or kicks.

Close Range

During close range engagements, combatants grab each other. Close range engagements also involve elbow strikes, knee strikes, and grappling.

2. Weapons of the Body

Hands and Arms

The hands, forearms, and elbows are the arm's individual weapons. The hands consist of several areas that can be used as weapons: fists, edges of hands, palms, and fingers.

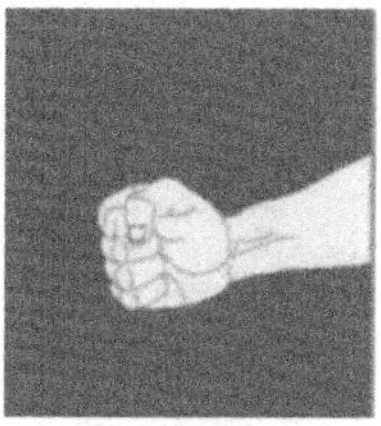

Fists. To minimize injury to the fists, Marines use their fists as weapons to target soft tissue areas such as the throat. The fists' striking surfaces are the first two knuckles of the hands or the meaty portions of the hands below the little fingers.

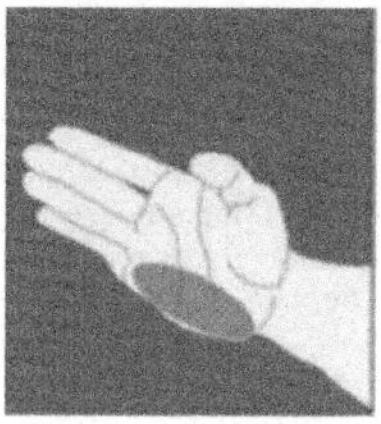

Edge of Hand. Marines use the edge of the hand (knife edge) as a weapon. Marines use the edge of the hand to strike soft tissue areas.

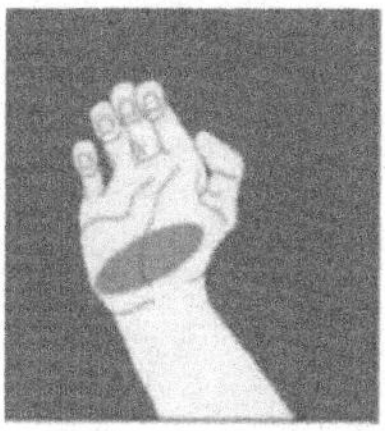

Palms. Because of the palm's padding, Marines use the heels of the palms to strike, parry, and/or block.

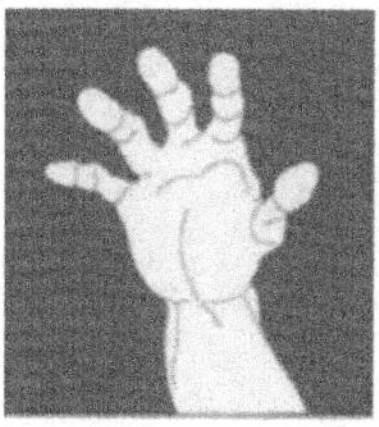

Fingers. Marines use the fingers to gouge, rip, and tear soft tissue areas (e.g., eyes, throat, groin).

Forearms. Marines use the forearms as a defensive tool to deflect or block attacks. Forearms can also be used as striking weapons to damage or break an opponent's joints and limbs. Marines sustain less self-injury when strikes are conducted with the forearms than when strikes are conducted with fists and fingers.

Elbows. Marines use the elbows as striking weapons. Because of the short distance needed to generate power, elbows are excellent weapons for striking during the close range of close combat.

Legs

The legs are more powerful than any other weapon of the body, and they are less prone to injury when striking. The feet are protected by boots and are the preferred choice for striking.

Feet. Marines use the balls of the feet, the insteps, and the toes to kick an opponent. Marines use the cutting edge of the heels and the heels to stomp on an opponent. Marines must be wearing boots when striking with the toes.

Knees. Like elbows, knees are excellent weapons in the close range of close combat. Knee strikes are most effective while fighting close to an opponent where kicks arc impractical. The opponent's groin area is an ideal target for the knee strike if he is standing upright. Knee strikes can deliver a devastating secondary attack to an opponent's face following an initial attack that caused him to bend at the waist.

3. Target Areas of the Body

During close combat, Marines strive to attack the accessible target areas of an opponent's body. The readily accessible areas will vary with each situation and throughout the engagement. The target areas are divided into five major groups: head, neck, torso, groin, and extremities. The figure below illustrates target areas of the body.

Head

The vulnerable regions of the head are the eyes, temple, nose, ears, and jaw. Massive damage to the head kills an opponent.

Eyes. The eyes are excellent targets because they are soft tissue areas that arc not protected by bone or muscle. Attacks to this area may cause the opponent to protect the area with his hands, allowing Marines to execute a secondary attack to other

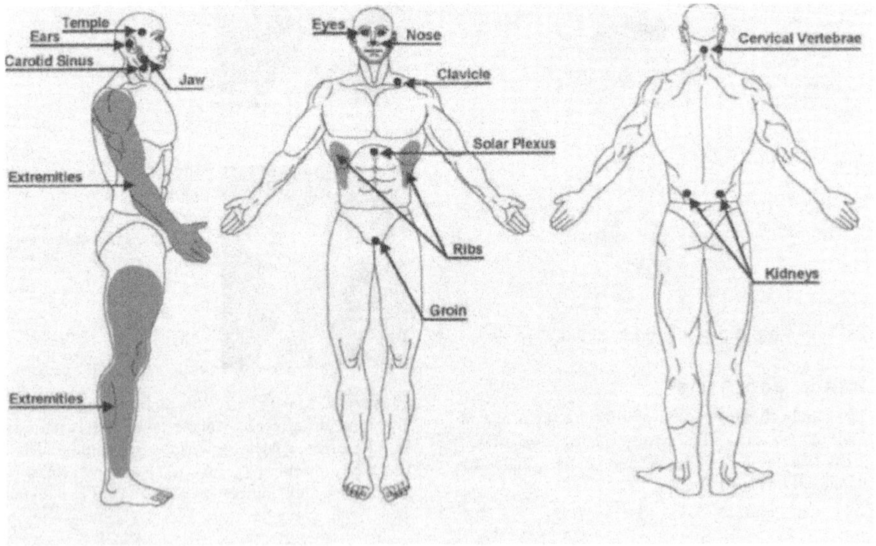

target areas while the opponent uses his hands to protect his eyes.

Temple. The temple is one of the most fragile areas of the head. Powerful strikes to the opponent's temple cause permanent damage and death.

Nose. The nose is very sensitive and easily broken. An attack to this area causes involuntary watering and closing of the opponent's eyes, rendering him vulnerable to secondary attacks. However, through training, individuals can condition themselves to withstand attacks to the nose. Therefore, any attack to the nose must be powerfully delivered.

Ears. Attacks to the ears may cause the eardrum to rupture. But this may not stop or even distract an opponent unless Marines powerfully deliver the strike.

Jaw. The jaw region, when struck forcefully, renders the opponent unconscious. Strikes to the jaw cause painful injuries to the teeth and surrounding tissues (e.g., lips, tongue), but the risk of self-injury is great unless Marines deliver strikes with a hard object such as a helmet, rifle butt, or boot heel.

Neck

The front of the neck, or throat area, is a soft tissue area that is not covered by natural protection. Damage to this region causes the opponent's trachea to swell, closing his airway, which can lead to death.

Carotid Sinus. The carotid sinus is located on both sides of the neck just below the jaw. Strikes to the carotid sinus restrict blood flow to the brain, causing loss of consciousness or death.

Cervical Vertebrae. The cervical vertebrae on the back of the neck, from the base of the skull to the top of the shoulders, contains the spinal cord, which is the nervous system's link to the brain. The weight of the head and the lack of large muscle mass allow damage to the cervical vertebrae and spinal cord. Excessive damage to this area causes pain, paralysis, or death.

Torso

Clavicle. The opponent's clavicle (or collar bone) can be easily fractured, causing immobilization of the arm.

Solar Plexus. Attacks to the opponent's solar plexus or center of the chest can knock the breath out of him and immobilize him.

Ribs. Damage to the opponent's ribs immobilizes him. It may also cause internal trauma.

Kidneys. Powerful attacks to the opponent's kidneys cause immobilization, permanent damage, or death.

Groin

The groin area is another soft tissue area not covered by natural protection. Any damage to this area causes the opponent to involuntarily protect his injured area, usually with his hands or legs. In male opponents, the scrotum is the main target since even a near miss causes severe pain, contraction of the lower abdominal muscles, deterioration of his stance, and possible internal trauma.

Extremities

Rarely will an attack to the opponent's extremities (arms and legs) cause death, but they are important target areas in close combat. Damage to an opponent's joints causes immobilization.

4. Pressure Points of the Body

There are nerves in the human body that, when pressure is applied or when they are struck, allow Marines to control a subject through pain compliance. Marines use pressure points to control an opponent when deadly force is not authorized. They also use pressure points to soften or distract an opponent so a lethal or nonlethal technique can be employed. The figure on page 1-5 illustrates the body's pressure points. Marines execute attacks to pressure points by—

: Rapidly kicking or striking pressure points.
: Slowly applying steady pressure to pressure points.

Infraorbital Nerve

The infraorbital nerve is just below the nose. Marines apply pressure to this nerve with an index finger to control the opponent.

Mastoid Process

The mastoid process is behind the base of the ear and beneath the edge of the jaw. Marines apply inward and upward pressure to this pressure point with the fingers to distract and control the opponent.

Jugular Notch

The jugular notch is at the base of the neck in the notch formed at the center of the clavicle. Marines apply pressure in a quick, stabbing motion with the index finger. Strikes to the jugular notch cause serious damage.

Brachial Plexus (Tie In)

The brachial plexus (tie in) is on the front of the shoulder at the joint. Strikes and pressure applied with the hand are effective on this nerve.

Radial Nerves

Radial nerves are on the inside of the forearms along the radius bones. Strikes and pressure applied with the hand to the radial nerve serve as a softening technique.

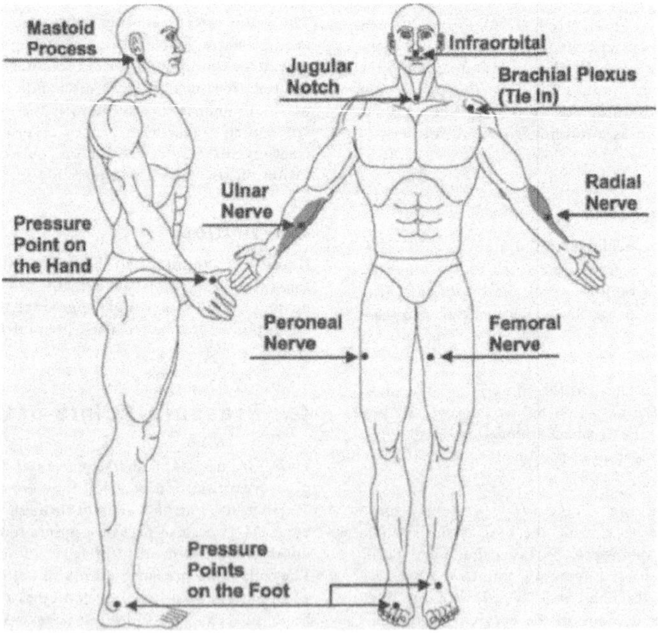

Ulnar Nerve

Ulnar nerves are on the outside of the forearms along the ulnar bones. Strikes and pressure applied with the hand to the ulnar nerve serve as a softening technique.

Pressure Point on the Hand

The hands contain a pressure point on the webbing between the thumbs and index fingers where the two bones of the fingers meet.

To force an opponent to soften or release his grip, Marines apply pressure with their index fingers to this pressure point or strike this pressure point with their fists.

Femoral Nerves

Femoral nerves are on the inside of the thighs along the femur bones. Strikes to the femoral nerve serve as a softening technique.

Peroneal Nerves

Peroneal nerves are on the outside of the thighs along the femur bones. Strikes to the peroneal nerve serve as a softening technique.

Pressure Points on the Feet

There are pressure points on the feet that, when pressure is applied or when they are struck, serve to soften or distract the opponent. Marines apply pressure with the toe, edge, or heel of their boots to the following points:

: The notch below the ball of the ankle.
: The top center of the foot, above the toes.
: The top of the foot where the leg and foot meet.

5. Basic Warrior Stance

Achieving a solid stance is critical to stability and movement throughout any close combat confrontation. The basic warrior stance provides the foundation for stability and movement that is needed to execute close combat techniques. To execute the basic warrior stance, Marines put their feet apart, hands up, elbows in, and chin down.

Feet Apart

Place feet shoulder-width apart.

Keep the head forward and eyes on the opponent, take a half step forward with the left foot, and pivot on the heels so the hips and shoulders are at approximately a 45-degree angle to the right.

Distribute body weight evenly on both legs. Bend the knees slightly.

Hands Up

Curl the fingers naturally into the palm of the hand. Position the thumb across the index and middle fingers. Do not clench the fists. Clenching the fists increases muscular tension in the forearms and decreases speed and reaction time.

Bring the hands up to the face at chin level, with the palms facing each other. Hold the fists up high enough to protect the head, but not so high that they block the field of vision. Ensure continuous eye contact with the opponent.

Elbows In

Tuck the elbows in close to protect the body.

Chin Down

Tuck the chin down to take advantage of the natural protection provided by the shoulders.

6. Angles of Approach and Movement

Marines use movement to control a confrontation and to retain a tactical advantage. Movement increases power and maximizes momentum. By moving around the opponent, Marines gain access to different target areas of the opponent's body. Movement allows Marines to use different weapons of the body and different close combat techniques to attack specific target areas.

Angles of Approach

Marines move anywhere within a 360-degree circle around the opponent to gain a tactical advantage. This circle provides access to different target areas of the opponent's body.

When facing an opponent, Marines move in a 45-degree angle to either side of the opponent. Moving at a 45-degree angle avoids an opponent's strike and puts Marines in the best position to attack the opponent. Marines should avoid being directly in front of an opponent during a confrontation. If a Marine is directly in front of an opponent, the opponent can rely on his forward momentum and linear power to seize the tactical advantage.

Movement

Marines must know how to move in all directions while maintaining the basic warrior stance. During any movement, the legs or feet should not be crossed. Once a movement is completed, the basic warrior stance should be resumed. Maintaining the basic warrior stance protects Marines and puts them in the proper position to launch an attack against an opponent.

Note: Before body movement begins. Marines turn their heads quickly to the new direction. The faster the head turns, the faster the body moves, and the quicker Marines attain visual contact with the opponent.

Forward to the Left. To move forward to the left from the basic warrior stance, Marines-

: Move the left foot forward at a 45-degree angle from the body (approximately 12 to 15 inches), keeping the toe pointed toward the opponent.

Bring the right foot behind the left foot as soon as the left foot is in place. This returns Marines to the basic warrior stance.

Forward to the Right. To move forward to the right from the basic warrior stance, Marines-

Move the right foot forward at a 45-degree angle from the body (approximately 12 to 15 inches).

Bring the left foot, toe pointing toward the opponent, in front of the right foot as soon as the right foot is in place. This returns Marines to the basic warrior stance.

Backward to the Left. To move backward to the left from the basic warrior stance, Marines execute the forward movement in reverse. Marines-

Move the left foot backward at a 45-degree angle from the body (approximately 12 to 15 inches), keeping the toe pointed toward the opponent.

Bring the right foot behind the left foot as soon as the left foot is in place. This returns Marines to the basic warrior stance.

Backward to the Right. To move backward to the right from the basic warrior stance, Marines execute the forward movement in reverse. Marines-

Move the right foot backward at a 45-degree angle from the body (approximately 12 to 15 inches).

Bring the left foot, toe pointing toward the opponent, in front of the right foot as soon as the right foot is in place. This returns Marines to the basic warrior stance.

7. Balance and Off-Balancing

Balance

In any close combat situation, Marines must strive to maintain balance. The last place to be in a close combat situation is on the ground. Marines must maintain a strong base and a low center of balance, their feet must be a shoulder-width apart, and they must stay on their toes to enable quick movement.

Off-balancing

Marines use off-balancing techniques to control an opponent. These techniques are used to throw an opponent to the ground while Marines remain standing, or they are used to put Marines in a position for an offensive attack.

Off-balancing techniques use the opponent's momentum to move or throw him. For example, if the opponent is charging a Marine, the Marine pulls the opponent to drive him to the ground. Likewise, if the opponent is pulling a Marine, the Marine pushes the opponent to drive him to the ground.

Off-balancing techniques also rely on the power generated by the opponent. For example, during combat a Marine may be tired or outnumbered. Depending on the generated energy and momentum of the opponent, the Marine employs off-balancing techniques with very little effort and still provides effective results.

Because off-balancing techniques rely on the momentum and power generated by the opponent, these techniques are particularly effective for Marines who may be outsized by their opponent or lack their opponent's strength.

Angles of Off-balancing

There are eight angles or directions in which an opponent can be off-balanced: forward, rear, right, left, forward right, forward left, rear right, and rear left.

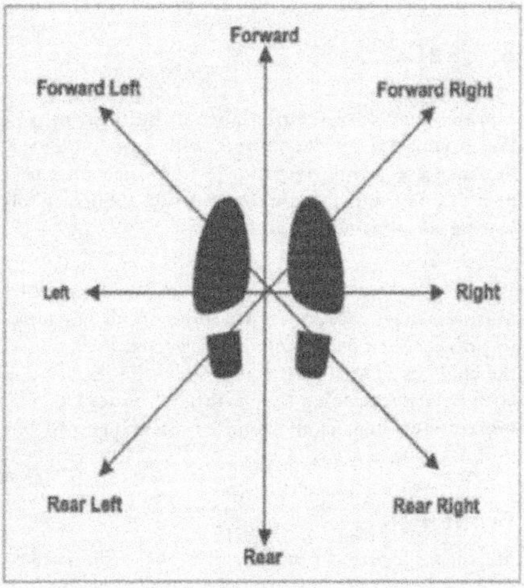

*Note: The angles correspond to the Marine's perspective, not the opponent's. Forward, rear, right, and left are straight angles. Forward right, forward left, rear right, and rear left are considered quadrants that are at a **45**-degree angle in either direction to the front or rear.*

Off-balancing Techniques

Marines off-balance an opponent by pushing, pulling, or bumping the opponent with their hands, arms, or bodies.

To pull, Marines grab an opponent with their hands and drive him forcefully to one of the rear quadrants or to the right or left.

To push, Marines grab the opponent with their hands and drive him forcefully into one of the front quadrants or to the right or left.

Marines execute bumping in the same manner as pushing, but use their shoulders, hips, and legs instead of their hands to off-balance the opponent.

8. Falls

Marines may lose their balance or be thrown to the ground during encounters with an opponent. Marines use falling techniques to absorb the impact of a fall and to quickly return to their feet following an opponent's attack.

Whether falling or being thrown by an opponent, Marines strive to reduce the force of the impact, to prevent serious personal injury, and to increase the chances of survival. Falling techniques use the body's large muscles (back, thighs, buttocks) to protect vital organs and bones from injury and immobilization.

Front Fall

Marines execute a front fall to break a fall on the front. To execute the front fall, Marines-

Bend the elbows and place the palms facing out in a position to spread and absorb the impact of the fall.

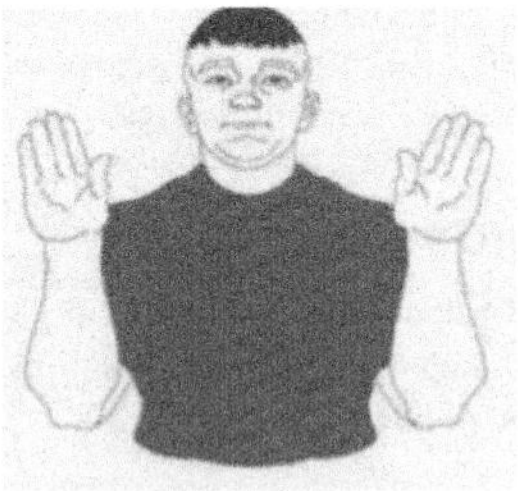

Fall forward, breaking the fall with the forearms and palms. The forearms and hands, down to the fingertips, should strike the ground simultaneously.

Offer resistance with the forearms and hands to keep the head raised off the ground.

Side Fall

Marines execute a side fall to break a fall on the side. To execute the side fall, Marines-

Bring the right arm across the body so the hand is next to the left shoulder with the palm facing inboard.

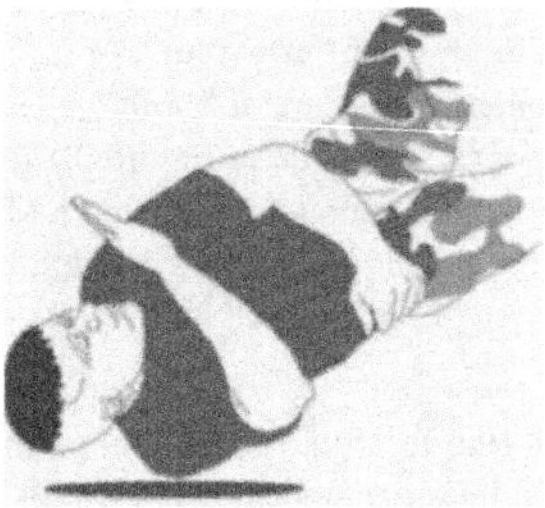

Fall to the side, breaking the fall with the right arm by slapping the ground and making contact from the shoulder or forearm down to the hand. At the same time, tuck the chin and keep the head raised off the ground. The chin should be tucked to the chest at all times to prevent whiplash.

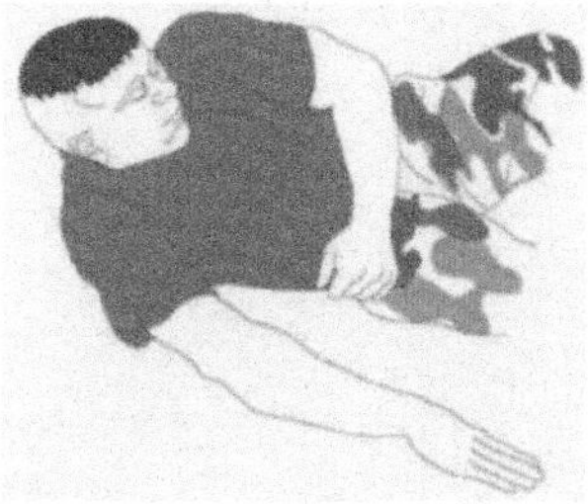

Stretch the right leg out to make contact with the ground and to distribute and absorb the impact. Bend the left leg, allowing the foot to make contact with the ground.

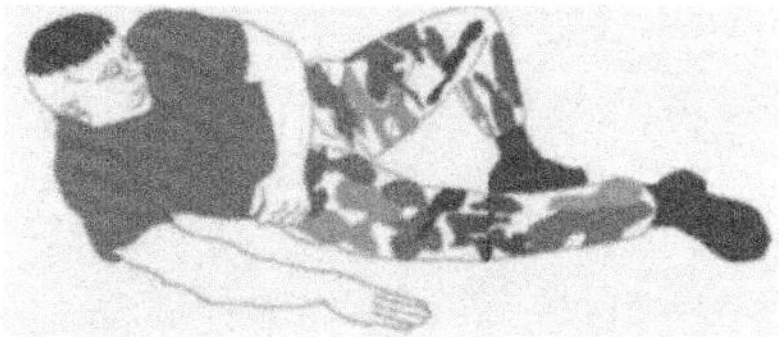

Back Fall

Marines execute a back fall to break the fall when being thrown or falling backward. To execute the backfall, Marines-

Cross the hands in front of the chest and tuck the chin.

Fall backward and slap the ground with the forearms and hands to absorb the impact of the fall and keep the head off the ground.

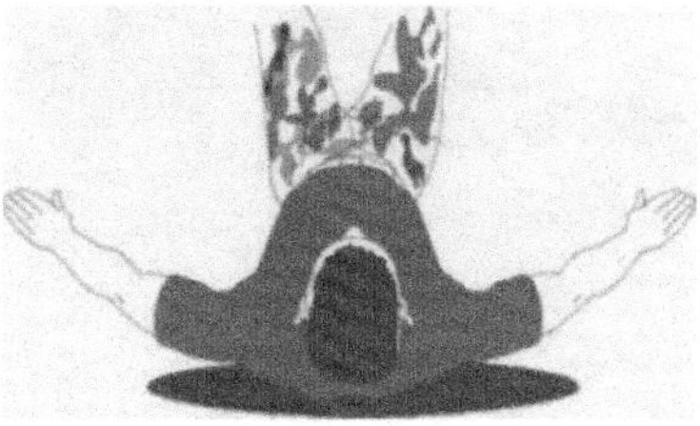

Forward Shoulder Roll

Marines use the forward shoulder roll to break a fall from an opponent's attack and to use the momentum of the fall to get back on their feet quickly. Ideally, Marines execute the forward shoulder roll to a standing position so they can continue fighting. To execute the forward shoulder roll, Marines-

Contact the ground with the back of the right forearm and upper arm. Tuck the chin into the chest.

Roll onto the right shoulder, rolling diagonally across the back to land on the left hip.

Slap the ground with the left arm, absorbing the impact from the shoulder to the hand, palm down.

Keep the left leg straight to absorb as much of the impact as possible. The right leg is bent and the foot hits flat on the ground.

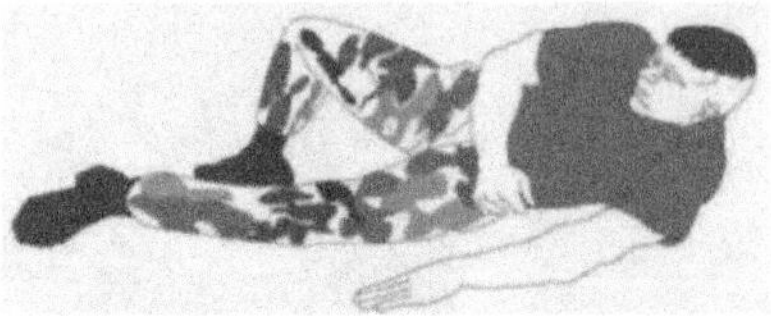

Bend the left leg upon impact to push off with the left knee and leg to a squatting and then a standing position. Forward momentum should carry the Marine to a standing position.

CHAPTER 2

LETHAL AND NONLETHAL WEAPONS TECHNIQUES

This chapter describes all techniques for a right-handed person. However, all techniques can be executed from either side.

In drawings, the Marine is depicted in woodland camouflage utilities; the opponent is depicted without camouflage. In photographs, the Marine is depicted in woodland camouflage utilities; the opponent is depicted in desert camouflage utilities.

1. Bayonet Techniques

WARNING

During training, Marines must have bayonets sheathed. Marines use bayonet dummies to practice bayonet techniques. When practicing offensive and defensive bayonet techniques student-on-student, Marines use pugil sticks.

All Marines armed with a rifle carry a bayonet. The bayonet is an effective weapon if Marines are properly trained in offensive and defensive bayonet techniques. An offensive attack, such as a thrust, is a devastating attack that can quickly end a fight. Defensive techniques, such as the block and parry, can deter the opponent's attack and allow Marines to regain the initiative. Through proper training, Marines develop the courage and confidence required to effectively use a bayonet to protect themselves and destroy the enemy. In situations where friendly and enemy troops are closely mingled and rifle fire and grenades are impractical, the bayonet becomes the weapon of choice.

Holding the Rifle

To execute bayonet techniques, Marines hold the rifle in a modified basic warrior stance. All movement begins and ends with the basic warrior stance. To hold the rifle, Marines—

Use an overhanded grasp to grab the small of the rifle's stock. Use an underhanded grasp to grab the handguards of the rifle.

Lock the buttstock of the rifle against the hip with the right forearm.

Orient the blade end of the rifle toward the opponent.

Offensive Bayonet Techniques

Straight Thrust. Marines use the straight thrust to disable or kill an opponent. It is the most deadly offensive technique because it causes the most trauma to an opponent. Target areas are the opponent' s throat, groin, or face. The opponent's chest and stomach are also excellent target areas if not protected by body armor or combat equipment. To execute the straight thrust, Marines—

Lift the left leg and lunge forward off the ball of the right foot while thrusting the blade end of the weapon forward, directly toward the opponent.

Retract the weapon and return to the basic warrior stance.

Slash. Marines use the slash to kill an opponent or to create an opening in his defense. The target area is the opponent's neck. To execute the slash, Marines—

Extend the left hand back toward the left shoulder.

Thrust the left hand forward and swing it to the right, bring the right hand back toward the hip, and turn the cutting edge of the blade toward the opponent's neck. The movement is a slashing motion so the blade cuts across the opponent's neck.

Horizontal Buttstroke. Marines use the horizontal buttstroke to weaken an opponent's defenses, to cause serious injury, or to set him up for a killing blow. Target areas are the opponent's head, neck, and legs. To execute the horizontal buttstroke, Marines—

Step forward with the right foot and drive the right hand forward. Rotate the hips and shoulders into the strike. Move the left hand back toward the left shoulder.

Strike the opponent with the butt of the weapon.

Vertical Buttstroke. Marines use the vertical buttstroke to weaken an opponent's defenses, to cause serious injury, or to set him up for a killing blow. Target areas are the opponent's groin and face. To execute the vertical buttstroke, Marines—

Step forward with the right foot and drive the right hand straight up.

Pull the left hand back over the left shoulder.

Strike the opponent with the butt of the weapon.

Smash. Marines use the smash as a follow-up technique to the horizontal or vertical buttstroke, particularly if they missed the target. The target area is the opponent's head. To execute the smash following a buttstroke, Marines-

Step forward with the right foot and place the blade end of the weapon over the left shoulder and elevate the right elbow above the shoulders.

Drive the arms straight forward, striking the opponent with the butt of the weapon.

Defensive Bayonet Techniques

Parry. Marines use a parry as a defensive technique to redirect or deflect an attack. A parry is a slight redirection of a linear attack by an opponent; e.g., a straight thrust or a smash. To execute the parry, Marines-

Use the bayonet end of the rifle to redirect the barrel or bayonet of the opponent's weapon.

Lock the weapon against the hip with the right forearm.

Rotate to the right or left, moving the bayonet end of the rifle to parry the opponent's attack. Rotation is generated from the hips.

Redirect or guide the opponent's weapon away from the body by exerting pressure against the opponent' s weapon.

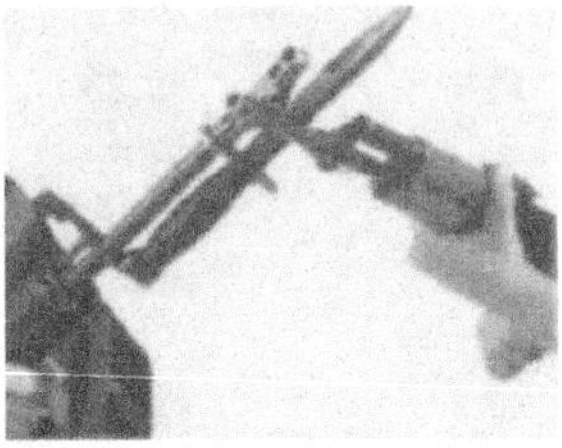

High Block. Marines execute a high block against a vertical attack coming from high to low. To execute the high block, Marines—

Thrust the arms up forcefully at approximately a 45-degree angle from the body. The weapon should be over the top of the head and parallel to the ground. The elbows are bent, but there is enough muscular tension in the arms to absorb the impact and deter the attack.

Low Block. Marines execute the low block

against a vertical attack coming from low to high. To execute the low block, Marines—

Thrust the arms down forcefully at approximately a 45-degree angle from the body. The weapon

should be at or below the waist and parallel to the ground. The elbows are bent, but there is enough muscular tension in the arms to absorb the impact and deter the attack.

Left and Right Block. Marines execute a left or right block against a horizontal buttstroke or a slash. To execute the left or right block, Marines-

Thrust the arms forcefully to the right or left, holding the rifle vertically in the direction of the attack. The elbows are bent, but there is enough muscular tension in the arms to absorb the impact and deter the attack.

Counter Action Following the Block. After deflecting an opponent's attack with a block, Marines counter with a slash or a horizontal buttstroke to regain the initiative. However, the objective in any bayonet fight is to thrust forward with the blade end of the weapon to immediately end the fight.

Group Strategy

On occasion, Marines may engage an opponent as a member of a group or numerous opponents by one's self or as a member of a group. By combining bayonet fighting movements and simple strategies, Marines can effectively overcome their opponent or opponents.

Offensive Strategy: Two Against One. If two bayonet fighters engage one opponent, the fighters advance together.

Fighter 1 engages the opponent while fighter 2 swiftly and aggressively attacks the opponent's exposed flank and destroys the opponent.

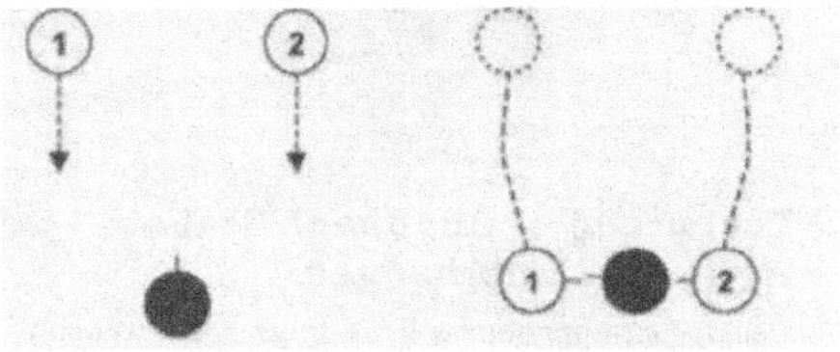

Offensive Strategy: Three Against Two. If three bayonet fighters engage two opponents, the fighters advance together keeping their opponents to the inside.

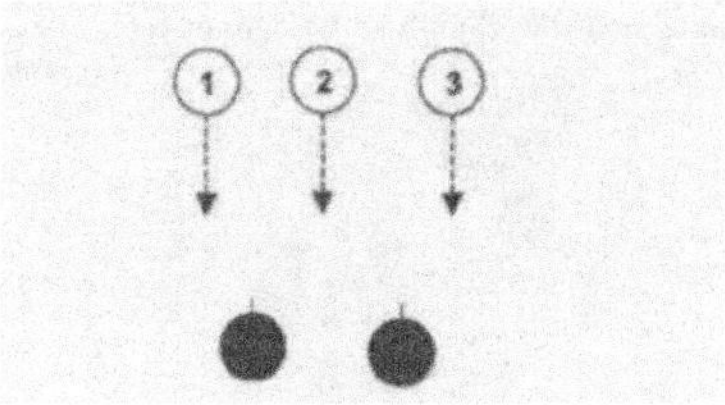

Fighters 1 and 3 engage opponents. Fighter 2 attacks the opponent's exposed flank engaged by fighter 1 and destroys the opponent.

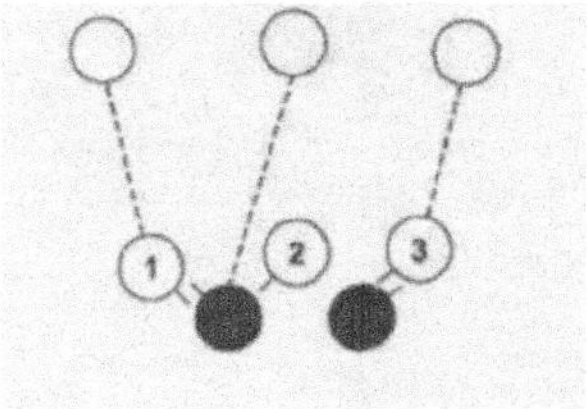

Fighters 1 and 2 turn and attack the exposed flank of the opponent engaged by fighter 3 and destroy the opponent.

Defensive Strategy: One Against Two. If a fighter is attacked by two opponents, the fighter immediately positions himself at the flank of the nearest opponent and keeps that opponent between himself and the other opponent.

Using the first opponent's body as a shield against the second opponent, the fighter destroys the first opponent quickly before the second opponent moves to assist.

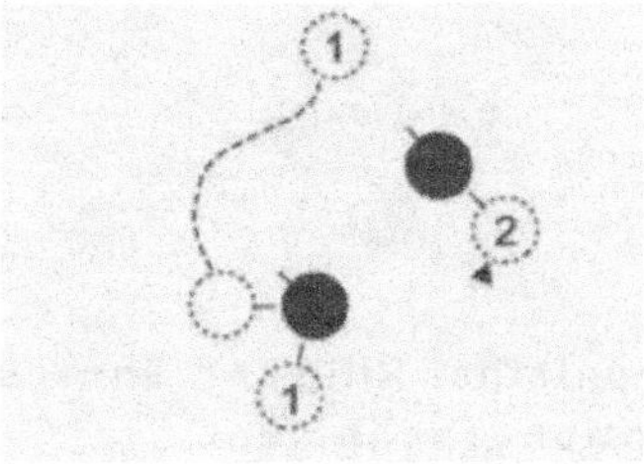

Then, the fighter engages and destroys the second opponent.

Defensive Strategy: Two Against Three. If two fighters arc attacked by three opponents, the fighters immediately move to the opponent's flanks.

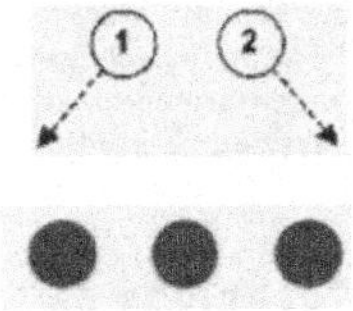

Fighters 1 and 2 quickly attack and destroy their opponents before the third opponent closes in.

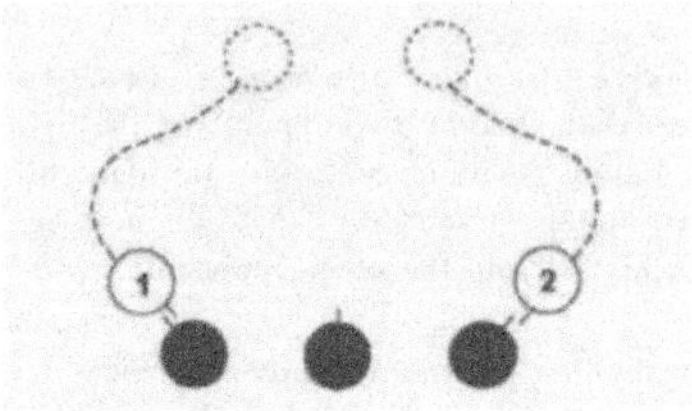

Fighter 1 engages the third opponent while fighter 2 attacks the opponent's exposed flank and destroys the opponent.

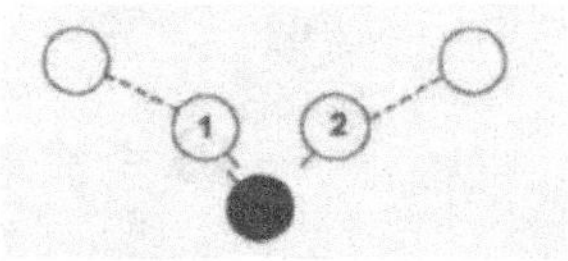

2. Nonlethal Rifle and Shotgun Retention Techniques

Most Marines are armed with the M16A2 service rifle. Marines are taught to keep their weapon with them at all times. Marines must be constantly alert to their surroundings and the people moving in and around their environment. Marines may be confronted by an individual who tries to take theii weapons. If this happens, Marines should not struggle with the individual. To retain positive control of their weapons, Marines must understand and apply weapons retention techniques, otherwise known as armed manipulation. The fol lowing techniques can be used with either the rifle or the shotgun.

Blocking Technique

To execute a blocking technique, Marines-

Stand in a defensive posture.

Use the weapon to block the opponent by thrusting it out firmly, with the elbows bent. Do not try to hit the opponent with the rifle. The rifle is used as a barrier.

Technique if Opponent Grabs Weapon Underhanded

Usually an opponent will try to grab the weapon or block it as an instinctive action. If the opponent uses an underhand grab to seize the handguards of the weapon, Marines—

Trap the opponent's closest finger(s) above the knuckle with the thumb so he cannot release his grip-

Apply bone pressure on the opponent's finger to initiate pain compliance.

Rotate the barrel of the weapon up or down quickly while maintaining pressure on the opponent's hand. At the same time, quickly pivot to off-balance the opponent.

Technique if Opponent Grabs Weapon Overhanded

If the opponent uses an overhand grab to seize the weapon overhanded, Marines—

Trap the opponent's finger to hold his hand in place.

Rotate the barrel to place it across the opponent's forearm and apply downward pressure. This action is similar to an armbar.

Technique if the Opponent Grabs the Muzzle

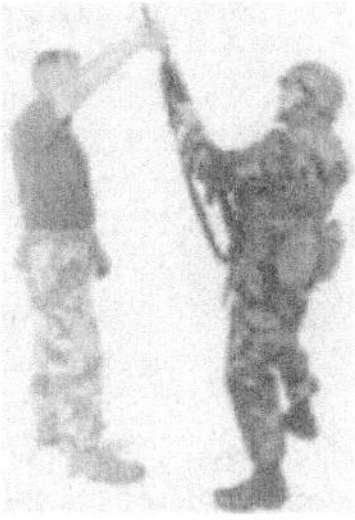

If the opponent grabs the muzzle of the weapon, Marines—

Rotate the muzzle quickly in a circle motion, Slash downward with the muzzle to release the opponent's grip.

Butt Strikes

Strikes with the butt of the weapon control or ward off an attacker. During any of the retention techniques, Marines use the heel or cutting edge of the weapon to deliver butt strikes to the inside or the outside of the opponent's thigh. The inside butt strike targets the femoral nerve as illustrated in the photo to the left.

The outside butt strike targets the peroneal nerve as illustrated in the photo to the left. Strikes can be made to the outside or inside of the thighs. If a strike to one side of the thigh misses, Marines follow back through with the butt of the weapon on the other side of the thigh.

Off-balancing Techniques

Marines apply off-balancing techniques to throw an opponent to the ground and retain possession of the weapon.

If the opponent grabs the weapon and pushes, Marines should not push on the weapon. They should—

Move with the momentum and movement of the

opponent by pivoting in the direction of the movement by stepping back.

Throw the opponent to the ground with a quick jerking movement by lowering the mu/./le and swinging the butt of the weapon.

If the opponent grabs the weapon and pulls, Marines—

Step on the opponent's foot and push forward to off-balance him and drive him to the ground.

Sweep the opponent's feet out from under him by hooking his leg with the leg and kicking backward.

3. Nonlethal Handgun Retention Techniques

Many Marines are armed with the M9 service pistol. Marines must keep their weapons in their possession at all times. Marines must be constantly alert to their surroundings and the people moving in and around the environment. Marines may be confronted by an individual who tries to take their weapons. To retain positive control of the weapon, Marines must understand and apply handgun retention techniques.

Blocking Technique

Marines perform the following blocking techniques if an opponent attempts to grab their pistol in the holster. Marines—

Place the body between the weapon and the opponent by immediately pivoting so the weapon is away from the opponent.

Step back and away from the opponent while placing the hand on the pistol grip.

Extend the left hand and block, deflect, or strike the opponent's arm.

Armbar Technique

Marines use the armbar technique when an opponent uses his right hand to grab their pistol in the holster. To execute the armbar technique, Marines—

Trap the opponent's right hand by grasping the opponent's wrist or hand with the right hand and applying pressure against the body.

Step back with the right foot and pivot sharply to the right to be next to the opponent. Always pivot in a direction that keeps the weapon away from the opponent.

Straighten the opponent's arm to apply an armbar. The arm should be straight across the torso.

Continue pivoting to the right while pulling back on the opponent's shoulder. This action may break the opponent's arm.

Place the left hand across the opponent's face and apply pressure back and down to take the oppo-

nent to the ground. Pressure applied beneath the nose or on the trachea is equally effective.

Wristlock Technique

Marines use the wristlock technique when an opponent grabs their pistol while it is in the holster with his right hand. To execute the w ristlock technique, Marines—

Grasp the opponent's wrist or hand with the right hand and apply pressure against the body.

Step back and away from the opponent and pivot to the right so the weapon is away from the opponent. Always pivot in a direction that keeps the weapon away from the opponent.

Reach over the opponent's arm with the left hand and grab his hand, applying pressure against his arm with the left forearm. Execute a wristlock.

Incorporate the second hand into the wristlock and, stepping back with the left foot, pivot to the left.

Execute a two-handed wristlock by exerting downward pressure with the thumb and rotating his hand to the left.

Continue to pivot to off-balance the opponent and drive him to the ground.

Note: If the opponent grabs the pistol with his left hand, Marines execute the wristlock with one hand and step in toward the opponent, rather than

away from the opponent.

Softening Techniques

Handgun retention techniques use softening techniques applied to pressure points. Bone pressure

and strikes with the hands (i.e., hammer fist), knees, and feet are also effective softening techniques.

Pressure Points. Marines use pressure point techniques to get the opponent to loosen his grip. Marines use their finger tips to apply pressure to the webbing between the index finger and thumb, the jugular notch, and the brachial plexus tie in. The following figure illustrates pressure applied to the brachial plexus tie in.

Bone Pressure. Bone pressure is the application of pressure on a bone against a hard object to initiate pain compliance. To apply bone pressure, Marines use their hand to trap the opponent's hand on the weapon. Marines apply a slow, steady pressure to the opponent's hand and fingers until his grip is softened or he releases his hold.

Strikes. If it is difficult to apply a retention technique, Marines employ strikes or kicks to force the opponent to loosen his grip. Strikes to the

eyes, the arms (radial nerve), or shoulder (brachial plexus tic in) soften the opponent's grip on the weapon.

Kicks and knee strikes to the peroneal nerve, the femoral nerve, or the groin arc effective because the opponent is typically unprepared to counter the strike.

Stomping on the top of the opponent's foot may distract him or loosen his grip on the weapon.

4. Firearm Disarmament Techniques

Marines use firearm disarmament techniques during a close-range confrontation if they are un-

armed and the opponent has a firearm (pistol). These techniques are equally effective if Marines are armed but do not have time to withdraw and present the weapon. The goal of firearm disarmament techniques is to gain control of the situation

so Marines gain the tactical advantage. The goal is not necessarily to get control of the opponent's weapon.

Pistol to the Front

This technique is used when Marines are unarmed and the opponent has a pistol pointing at their front (e.g., head, chest). The technique is the same if the opponent sticks the pistol under the Marine's chin. To execute the counter when an opponent is pointing a pistol toward the front of a Marine, Marines—

Place the hands close to the weapon, about chest high, palms out.

Use the left hand to grab the opponent's forearm and push the opponent's hand with the pistol away to clear the body from in front of the weapon. At the same time, rotate the right shoulder back to clear the body from the weapon.

Maintain control of the opponent's arm.

Grasp the weapon with the right hand by placing the thumb underneath the pistol and the fingers over top of the pistol.

Keep the right hand wrapped tightly around the muzzle and quickly rotate the pistol in the opponent's hand so the muzzle is facing the opponent.

Grasp and pull the opponent's wrist or forearm away from the body while rotating the weapon.

Rotate the weapon toward the opponent while pulling it up and back and out of the opponent's grasp.

Pistol to the Rear

This technique is performed when Marines are unarmed and the opponent has a pistol pointing to the back of the Marine's head. To execute the counter when the opponent is pointing a pistol to the rear, Marines—

Place the hands close together about chest high, palms out.

Step back with the left foot, pivoting on the right foot so the side is against the opponent's front. This action clears the body from the weapon's line of fire. Keep the left hand up.

Pivot on the left foot to face the opponent and, at the same time, raise the left elbow and reach over the top of the opponent's arm with the left arm.

Wrap the left arm tightly around the opponent's arm above his elbow to control it.

Push on the opponent's shoulder with the right hand while pulling up with the left arm to achieve an armbar. This action releases the opponent's grip on the weapon. If necessary, the opponent can be taken to the ground with a leg sweep.

Note: To execute this technique, the weapon must be close to or touching the rear of the Marine. If the weapon is too far away from the body, this technique would be difficult to execute or it would be ineffective.

CHAPTER 3

HAND-HELD WEAPONS

This chapter describes all techniques for a right-handed person. However, all techniques can be executed from either side.

In drawings, the Marine is depicted in woodland camouflage utilities; the opponent is depicted without camouflage. In photographs, the Marine is depicted in woodland camouflage utilities; the opponent is depicted in desert camouflage utilities.

Marines must know how to defend against attacks when an opponent is either unarmed or armed with a held-held weapon. This chapter addresses the combative use of knives, specific weapons of opportunity, and sticks. However, virtually anything can be used as a hand-held weapon.

1. Fundamentals of Knife Fighting

Marines must be trained in knife fighting techniques. Marines experienced in offensive knife techniques can cause enough damage and massive trauma to stop an opponent. When engaged against each other, experienced knife fighters employ various maneuvers and techniques that are specific to knife fighting. Seldom, if ever, will Marines engage an opponent in a classical knife fight.

Note: When armed with a rifle, Marines are issued a bayonet. When armed with a pistol, Marine are issued a combat knife.

Angles of Attack

There are six angles from which an attack with a knife can be launched:

- Vertical strike coming straight down on an opponent.
- Forward diagonal strike coming in at a 45-degree angle to the opponent.
- Reverse diagonal strike coming in at a 45-degree angle to the opponent.
- Forward horizontal strike coming in parallel to the ground.
- Reverse horizontal strike coming in parallel to the ground.
- Forward thrust coming in a straight line to the opponent.

Target Areas of the Body

During any confrontation, the parts of the opponent's body that are exposed or readily accessible will vary. The goal in a knife fight is to attack the body's soft, vital target areas that are readily accessible (e.g, the face, the sides and front of the neck, the lower abdomen [or groin]).

Neck. Carotid arteries, located on either side of the neck, are good target areas because they are not covered by body armor or natural protection.

Lower Abdomen (or Groin). The lower abdomen (or groin region) is a good target area because it is not covered by body armor.

Heart. The heart, if not covered by body armor, is an excellent target which, if struck, can prove fatal in a matter of seconds or minutes.

Secondary Targets. There are secondary target areas that will cause substantial bleeding if an artery is severed. These target areas are not immediately fatal, but can become fatal if left unattended. Attacks to—

The legs can cause a great deal of trauma and prove fatal. For example, the femoral artery located on the inside of the thigh is a large artery which, if cut, will cause extensive blood loss.
The brachial artery, located between the biceps and triceps on the inside of the arm, can cause extensive bleeding and damage.
The arm's radial and ulnar nerves can cause extensive bleeding and damage.

Movement

Marines can move anywhere within a 360-degree circle around the opponent. This allows accessibility to different target areas of the opponent's body. Marines should avoid being directly in front of an opponent because the opponent can rely on his forward momentum to seize the tactical advantage. If Marines face an opponent, movement is made in a 45-degree angle to either side of the opponent. This angle avoids an opponent's strike and places Marines in the best position to attack an opponent.

Wearing the Combat Knife

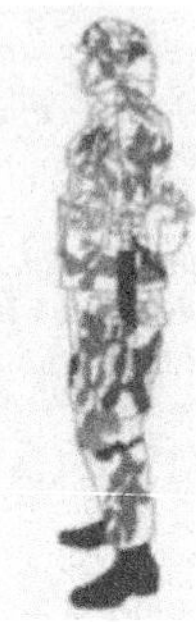

Marines must wear the combat knife where it is easily accessible and where it can best be retained. It is recommended the combat knife be worn on the weak side hip, blade down, sharp edge facing forward. Marines can place it behind the magazine pouch where it is easily accessible to them, but not easily grabbed by an opponent.

Grip

The grip on the knife should be natural. Marines grasp the knife's grip with the fingers wrapped around the grip naturally as it is pulled out of its sheath. This is commonly known as a hammer grip. The blade end of the knife is always facing the opponent.

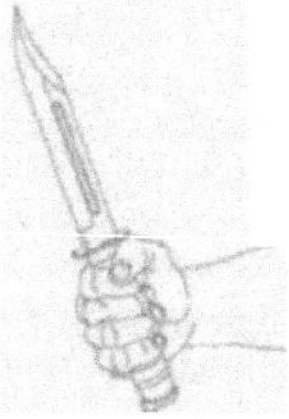

Stance

Marines use the basic warrior stance as the foundation for knife techniques. The left hand is a vertical shield that protects either the ribs or the head and neck. The right elbow is bent with the blade pointing forward toward the opponent's head. This position serves as an index point, where all techniques are initiated.

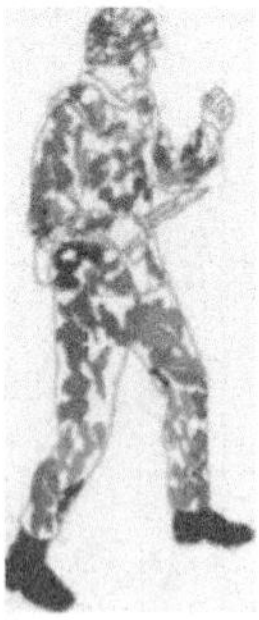

Principles of Knife Fighting

The following are key principles of knife fighting:

Execute movements with the knife blade within a box, shoulder-width across from the neck down to the waistline. The opponent has a greater chance of blocking an attack if the blade is brought in a wide, sweeping movement to the opponent.
Close with the opponent, coming straight to the target.
Move with the knife in straight lines.
Point the knife's blade tip forward and toward the opponent.
Apply full body weight and power in each of the knife techniques. Full body weight should be put into the attack in the direction of the blade's movement (slash or thrust).
Apply constant forward pressure with the body and blade to keep the opponent off-balanced.

2. Knife Fighting Techniques

Slashing Techniques

Marines use slashing techniques to close with an enemy. Slashing techniques distract the opponent or damage the opponent so Marines can close in. Typically, Marines target the opponent's limbs, but any portion of the body that is presented can become a target.

Vertical Slash Technique. The vertical slash follows a vertical line straight down through the target. To execute the vertical slash, Marines—

Thrust the right hand out and bring the weapon straight down on the opponent, continuing to drag the knife down through the opponent's body.

Maintain contact on the opponent's body with the blade of the knife.

Forward Slash Technique. The forward slash follows a straight line in a forehand stroke, across the target areas of either the neck (high diagonal slash) or abdominal region (low horizontal slash). To execute the forward slash, Marines-
Extend the right hand while simultaneously rotating the palm up until the knife blade makes contact with the opponent.

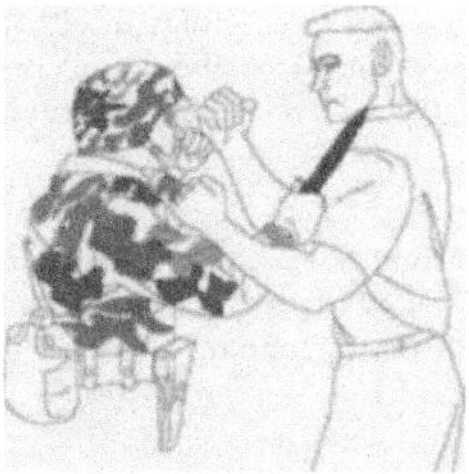

Snap or rotate the wrist through the slashing motion to maximize blade's contact with the opponent.

Drag the knife across the opponent's body, from right to left, in a forehand stroke. The movement ends with the forearm against the body and the knife at the left hip with its blade oriented toward the opponent.

Reverse Slash Technique. The reverse slash is a follow-up technique to a forward attack. It allows Marines both a secondary attack and the ability to resume the basic warrior stance. The reverse slash follows a straight line in a backhand stroke, across the target areas of either the neck (high diagonal slash) or abdominal region (low horizontal slash). To execute the reverse slash, Marines—

Extend the right hand while simultaneously rotating the palm down until the knife blade makes contact with the opponent.

Snap or rotate the wrist through the slashing motion to maximize the blade's contact with the opponent.

Drag the knife across the opponent's body, from left to right, in a backhand stroke. Maintain contact on the opponent's body with the blade of the knife.

Thrusting Techniques

The primary objective of knife fighting is to insert the blade into an opponent to cause extensive damage and trauma. This is done with a thrusting technique. Thrusting techniques are more effective than slashing techniques because of the damage they can inflict. However, Marines use slashing techniques to close with the enemy so that they are closer to the opponent, which allows them to use the thrusting technique.

Vertical Thrust. The thrusting motion follows a vertical line straight up through the target (low into the abdomen region or high into the neck). To execute the vertical thrust, Marines—

Thrust the right hand toward the target, inserting the knife blade straight into the opponent.

Pull the knife out of the opponent.

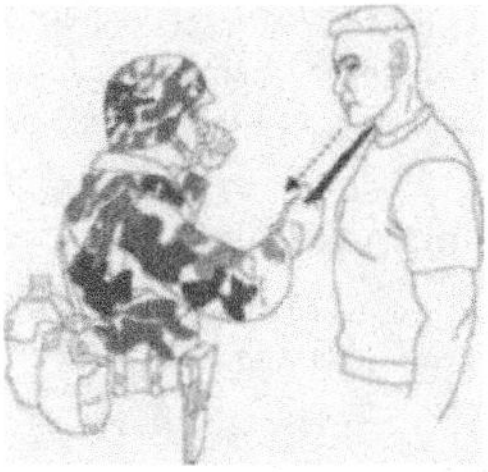

Forward Thrust. The forward thrust follows a straight line straight into the opponent's neck (high thrust) or abdominal region (low thrust). To execute the forward thrust, Marines—

Thrust the right hand, palm down, toward the target, inserting the knife blade straight into the opponent.

Rotate the palm up once the knife is inserted to twist the blade.

Drop the right elbow and bring the knife to the opposite side of the opponent's body from where it was inserted. At the same time, rotate the hips and shoulders downward to bring body weight to bear on the attack.

Reverse Thrust. The reverse thrust is a follow-up technique to a forward attack. It allows Marines both a secondary attack and the ability to resume the basic warrior stance. The reverse thrust follows a horizontal line straight into the opponent's neck (high thrust) or abdominal region (low thrust). To execute the reverse thrust, Marines-

Bend the right arm, crossing the arm to the left side of the body.

Thrust the right hand, palm up, toward the target, inserting the knife blade straight into the opponent.

Rotate the palm down to twist the blade once the knife is inserted.

Bring the knife to the opposite side of the opponent's body from where it was inserted.

3. Weapons of Opportunity

During an unarmed close combat situation, Marines use their bodies as weapons, but they should be ready and able to use anything around them as a weapon. For example, Marines could throw sand or liquid in an opponent's eyes to temporarily impair his vision or smash the opponent's head with a rock or helmet. Marines must use whatever means are available and do whatever it takes to take control of the situation and to win, or they face the possibility of losing their lives. Some weapons of opportunity are discussed in the following subparagraphs.

Entrenching Tool

An entrenching tool (E-tool) is commonly carried by Marines. It can be an excellent weapon, especially when sharpened. Marines can use the E-tool to block, slash, and thrust at an opponent.

Tent Pole and Pins

Marines can use tent poles and pins to block, strike, or thrust at an opponent.

Web Belt

Marines can stretch a web belt between their hands to block attacks by an opponent.

Battlefield Debris

Marines can use debris on the battlefield (e.g., sticks, glass, a sharp piece of metal) to cut, slash, or stab an opponent. They can also use other types of debris such as shovel or ax handles, boards, metal pipes, or broken rifles to strike an opponent or apply a choke.

Helmet

A helmet can be used to strike an opponent on an unprotected area like the head and face. Grasp the rim of the helmet and thrust the arms forward, striking the opponent with the top of the helmet.

4. Fundamentals of Combative Stick

On the battlefield, Marines must be ready and able to use anything as a weapon. They must learn and be able to use techniques that can be employed with most weapons of opportunity. Among these techniques are combative stick techniques. Combative stick techniques can be used with a stick, a club, a broken rifle, an E-tool, or even a web belt.

Angles of Attack

There are six angles from which an attack with a hand-held weapon can be launched:

- Vertical strike coming straight down on an opponent.
- Forward diagonal strike coming in at a 45-degree angle to the opponent.
- Reverse diagonal strike coming in at a 45-degree angle to the opponent.
- Forward horizontal strike coming in parallel to the ground.
- Reverse horizontal strike coming in parallel to the ground.
- Forward thrust coming in a straight line to the opponent.

Grip

Grasp the stick about 2 inches from its base.

Stance

The basic warrior stance serves as the foundation for combative stick techniques. The left hand becomes a vertical shield that protects the ribs or the head and neck. Depending on how heavy the weapon is, it should be held at a level approximately shoulder height.

Movement

Movement during combative stick techniques is the same as it is for other close combat techniques. Marines can move anywhere within a 360-degree circle around the opponent. This allows accessibility to different target areas of the opponent's body and gains a tactical advantage.

Marines should avoid being directly in front of an opponent because the opponent can rely on his forward momentum to seize the tactical advantage. If Marines face an opponent, movement is made in a 45-degree angle to either side of the opponent. This angle avoids an opponent's strike and places Marines in the best position to attack an opponent.

5. Combative Stick Techniques

Strikes

Strikes are intended to inflict as much damage on an opponent as possible. Striking techniques apply to a weapon of opportunity such as a stick, a tent pole, a club, a broken rifle, an E-tool, or a Pipe-

Vertical Strike. To execute the vertical strike, Marines—

Bend the right arm, extending the weapon over the back of the right shoulder.

Rotate the forearm straight down off the elbow to bring the weapon down on the opponent.

Rotate the hips and shoulders forcefully toward the opponent.

Follow through with the strike by allowing the weight of the weapon to go through the target area of the body.

Forward Strike. To execute a forward strike, Marines—

Step forward with the left foot in the direction of the strike.

Bend the right arm with the elbow extending out to the right and the weapon extended over the right shoulder.

Rotate the forearm to the right of the elbow to bring the weapon down onto the opponent. At the same time, forcefully rotate the hips and shoulders toward the opponent.

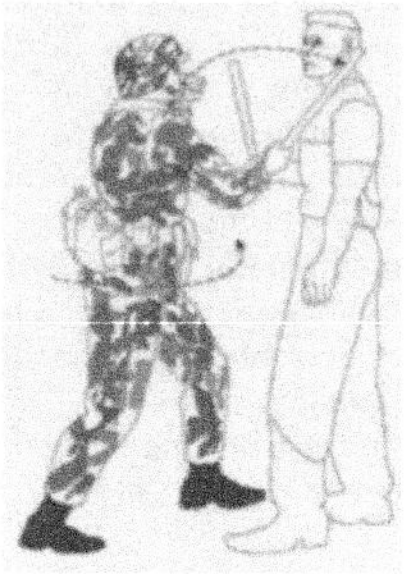

Follow through with the strike by allowing the weight of the weapon to go through the target area of the body.

Reverse Strike. The reverse strike is a follow-up technique to a forward strike. It allows Marines both a secondary attack and the ability to resume the basic warrior stance.To execute a reverse strike, Marines—

Step forward with the right foot in the direction of the strike.

Bend the right arm with the hand near the left shoulder. The weapon is extended over the left shoulder.

Rotate the forearm to the right of the elbow to bring the weapon down onto the opponent. At the same time, forcefully rotate the hips and shoulders toward the opponent.

Follow through with the strike by allowing the weight of the weapon to go through the target area of the body.

Forward Thrust. To execute the forward thrust, Marines—

Grasp the stick with the left hand, palm up, in a position where the stick can be controlled with two hands.

Lift the left leg and lunge forward off the ball of the right foot. At the same time, thrust the end of the weapon directly toward the opponent by thrusting both hands forward in a straight line.

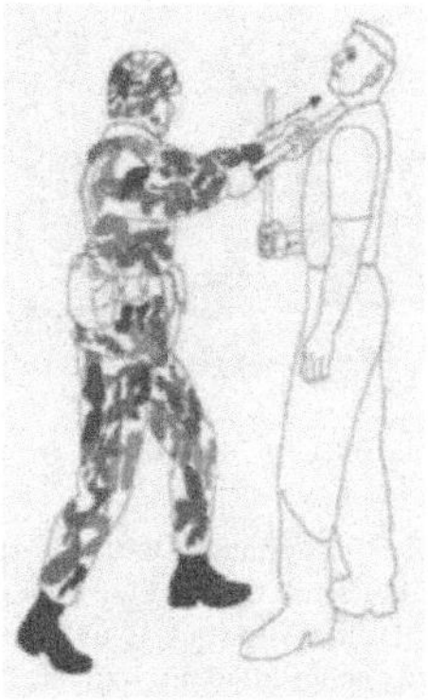

6. Blocking Techniques

A block is meant to deter or deflect an attack by an opponent. A block sets up Marines for a follow-on attack against the opponent. Blocks are executed by deflecting, rather than hitting or following through like a strike.

Blocks Against Unarmed Attacks

To block against an unarmed attack, Marines—

Step forward at a 45-degree angle with either the right or left foot. This moves the body out of the line of attack.

Raise the left arm and block the strike with the meaty portion of the forearm.

Employ the stick by one of two techniques:

Strike the opponent with the stick.
Use the stick to block on two points of contact. When the stick is used to block, it serves as an incidental strike. This technique is only used if Marines have closed with the opponent and are inside his strike.

Blocks Against Armed Attacks

Block for a Vertical Strike. To execute the block for a vertical strike, Marines—

Step forward with the left foot at a 45-degree angle to the left. This moves the body out of the line of attack.

Block on two points of contact to disperse the impact of the attack:

Block the opponent's stick by positioning the stick so it is perpendicular to the opponent's stick. If the stick is not perpendicular to the opponent's stick, the stick can slide through and make contact on the Marine.
Block the opponent's wrist or forearm with the back of the left forearm.

Note: Use the stick to block the opponent's arm if closer to the opponent. It is the same movement as blocking with the arm, except the opponent's arm is blocked with both the stick and the arm.

Block for a Forward Strike. To execute the block for a forward strike, Marines-

Step forward with the left foot at a 45-degree angle to the left. This moves the body out of the line of attack and inside the opponent's strike.

Block on two points of contact to disperse the impact of the attack:

- Block the opponent's wrist or forearm with the meaty portion of the back of the left forearm.
- Strike the opponent's attacking biceps with the stick.

Block for a Reverse Strike. To execute the block for a reverse strike, Marines—

Step forward at a 45-degree angle with the right foot to the right. This moves the body out of the line of attack and inside the opponent's strike.

Block on two points of contact to disperse the impact of the attack:

Block the opponent's stick by positioning the stick so it is perpendicular to the opponent's stick.

Block the opponent's forearm with the meaty portion of the left forearm.

Note: If closer to the opponent, block his triceps with the back of the left forearm and strike his forearm with the stick.

7. Unarmed Against Hand-Held Weapons

If Marines are engaged against an opponent with a knife, a stick, or some other weapon of opportunity, they must establish and maintain an offensive mindset, not a defensive mindset. Their survival depends on it. They cannot afford to think about getting cut or hurt.

Angles of Attack

Before Marines can learn to block or counter an attack with a hand-held weapon (e.g., knife, stick), they must know from what angle the opponent is attacking. There arc six angles from which an opponent will typically attack with a hand-held weapon:

Vertical strike coming in straight down on the Marine.
Forward diagonal strike coming in at a 45-degree angle to the Marine.
Reverse diagonal strike coming in at a 45-degree angle to the Marine.
Forward horizontal strike coming in parallel to the ground.
Reverse horizontal strike coming in parallel to the ground.
Forward thrust coming in a straight line to the Marine.

Blocks

If the opponent has a hand-held weapon (e.g., knife or stick), Marines parry the opponent's hand or arm to block the attack.

Basic Block. To execute the basic block technique, Marines-

Step forward at a 45-degree angle to move out of the line of the attack. Always step in the direction of the strike.

Thrust the forearms forward, hands up, against the opponent's attacking arm. Contact is made on the opponent's arm with the backs of the forearms.

Block for a Vertical Strike. To execute the block against a vertical strike, Marines—

Step forward with the left foot at a 45-degree angle to the left to move out of the line of attack.

Thrust the forearms forward, hands up, against the outside of the opponent's attacking arm.

Block for a Forward Strike. To execute the block against a forward diagonal or forward horizontal strike, Marines-

Step forward with the left fool inside the opponent's attacking arm.

Block the attack with both arms bent so the forearms make contact with the opponent's biceps and forearm.

Block for a Reverse Strike. To execute the block against a reverse diagonal or reverse horizontal strike, Marines—

Step forward with the right foot to the outside of the opponent's attacking arm.

Block the attack with both arms bent so the forearms make contact with the opponent's triceps and forearm.

Block for a Forward Thrust. To execute the block against a forward thrust, Marines—

Bend at the waist, move the hips backward, and jump backward with both feet to move away from the attack. This action is known as "hollowing out."

Hollow out and block the attack with the arms bent and hands together on top of the attacking arm.

Overlap the hands slightly so one thumb is on top of the other hand's index finger. The other thumb should be under the other hand's index finger.

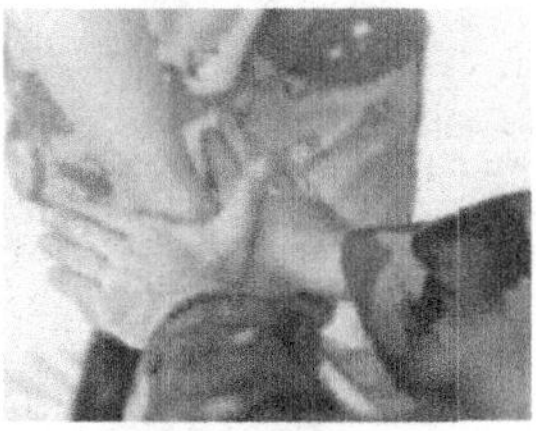

8. Counters to Hand-Held Weapon Attacks

Principles of Counters

A counter is used to control the situation to regain the tactical advantage and end the fight. Regardless of the type of weapon or angle of attack, the following principles apply to countering the attack with a hand-held weapon:

Move out of the line of attack. Movement is executed in a 45-degree angle forward to the left or right.

- Block the attack.

Note: The first two actions are taken simultaneously.

- Take control of the weapon by controlling the hand or arm that is holding the weapon. Never attempt to grab the opponent's weapon.
- Execute the appropriate follow-up to end the fight; e.g., strikes, joint manipulations, throws, or takedowns (see chap. 8). Marines should continue to attack the opponent until the fight ends.

Counter Techniques

There are two techniques that can be used to counter any armed attack: forward armbar counter and reverse armbar counter. These techniques can be used to counter a vertical attack, a forward diagonal strike, or a forward horizontal strike. With minor variations, the same techniques arc used to counter reverse strikes. A third technique, the bent armbar counter, is used to counter a vertical attack.

Forward Armbar Counter. To execute the forward armbar counter to an attack coming from a forward strike, Marines-

Step forward with the left foot inside the opponent' s attacking arm.

Block the attack with both arms bent so the forearms make contact with the opponent's biceps and forearm.

Slide the left arm over the opponent's forearm and wrap the arm tightly around his arm, trapping the opponent's attacking arm between the biceps and torso.

Place the right hand on the opponent's shoulder or upper arm to further control his arm and to effect an armbar.

Execute an armbar and continue to exert steady pressure against the arm.

Reverse Armbar Counter. To execute the reverse armbar counter to an attack coming from a forward strike, Marines-

Step forward with the left foot inside the opponent' s attacking arm.

Block the attack with both arms bent so the forearms make contact with the opponent's biceps and forearm.

Control the opponent's arm with the left hand and pivot to the right so the back is against the opponent's side. Immediately slide the right arm over the opponent's biceps and wrap the arm tightly around his arm, trapping the opponent's attacking arm between the biceps and torso.

Grasp the opponent's wrist with the left hand and twist his thumb away from the body.

Take the opponent to the ground with an armbar takedown if he does not drop the weapon from the wristlock.

Bent Armbar Counter. This counter is particularly effective against a vertical attack. To execute the bent armbar counter, Marines—

Step forward with the left foot inside the opponent's attacking arm.

Block the attack with both arms bent so the forearms make contact with the opponent's biceps and forearm.

Grasp the opponent's forearm with the left hand. At the same time, slide the right arm underneath the opponent's triceps and grasp the opponent's forearm or wrist with the right hand.

Apply downward pressure with the hands against the opponent's forearm to off-balance the opponent.

CHAPTER 4

STRIKES

This chapter describes all techniques for a right-lmndedperson. However, all techniques can be executed from either side.

In drawings, the Marine is depicted in woodland camouflage utilities; the opponent is depicted without camouflage. In photographs, the Marine is depicted in woodland camouflage utilities; the opponent is depicted in desert camouflage utilities.

Strikes are unarmed individual hitting techniques. Strikes use the hands, elbows, knees, feet, and, in some instances, other parts of the body as personal weapons. Marines must know how to execute strikes effectively. They must also know how to counter strikes from an opponent.

1. Principles of Punches

Muscle Relaxation

Muscle relaxation is crucial when executing punches. The natural tendency in a fight is to tense up, which results in rapid fatigue and decreased power generation. Marines who remain relaxed during a close combat situation generate greater speed, which results in greater generation of power. Relaxing the forearms generates speed and improves reaction time. At the point of impact, Marines clench the fist to cause damage to the opponent and avoid injury to the wrist and hand.

Weight Transfer

Weight transfer is necessary to generate power in a punch. Marines accomplish this by-

Rotating their hips and shoulders into the attack.

Moving their body mass straight forward or backward in a straight line.

Dropping their body weight into an opponent. Body mass can be transferred into an attack from high to low or from low to high.

Rapid Retraction

When Marines deliver a punch, rapid retraction of the fist is important. Once the hand has made contact with the target, Marines quickly return to the basic warrior stance. Rapid retraction-

Returns the hand and arm to the protection afforded by the basic warrior stance.

Prevents the opponent from being able to grab the hand or arm.

Permits the hand and arm to be "chambered" or "re-cocked" in preparation for delivering a subsequent punch.

Telegraphing

Telegraphing a strike occurs when body movements inform the opponent of the intention to launch an attack. Staying relaxed helps to reduce telegraphing.

Often, an untrained fighter telegraphs his intention to attack by drawing his hand back in view of his opponent, changing facial expression, tensing neck muscles, or twitching. These movements, however small, immediately indicate an attack is about to be delivered. If the opponent is a trained fighter, he may be able to evade or counter the attack. If the opponent is an untrained fighter, he may still be able to minimize the effect of an attack.

2. Punches

Punches may be thrown during any hand-to-hand confrontation. Most people resort to punching because it is a natural reaction to a threat. The purpose of a punch is to stun the opponent or to set him up for a follow-up finishing technique. However, punches should only be executed to the soft tissue areas of an opponent. A correctly delivered punch maximizes the damage to an opponent while minimizing the risk of injury to Marines.

Basic Fist

Punches are executed using the basic fist. To make the basic fist, the fingers are curled naturally into the palm of the hand and the thumb is placed across the index and middle fingers. Do not clench the fist until movement has begun. This reduces muscular tension in the forearm and increases speed and reaction time. Just before impact, Marines exert muscular tension on the hand and forearm to maximize damage to the opponent and reduce their chances of injury. Contact should be made with the knuckles of the index and middle fingers.

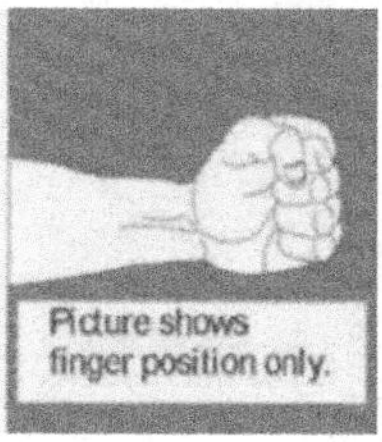

When striking with the basic fist, Marines must keep the hand straight, or in line, with the wrist to avoid injury to the wrist.

Lead Hand Punch

The lead hand punch is a snapping straight punch executed by the forward or lead hand. It is a fast punch designed to keep the opponent away and to set up an attack. A lead hand punch conceals movement and allows Marines to get close to the opponent. Lead hand punches should strike soft tissue areas, if possible. To execute the lead hand punch, Marines-

Snap the lead hand out to nearly full extension, while rotating the palm to the ground.

Contact the opponent with the first two knuckles of the fist.

Retract the hand immediately, resuming the basic warrior stance.

Rear Hand Punch

The rear hand punch is a snapping punch executed by the rear (right) hand. It is a power punch designed to inflict maximum damage on the opponent. Its power comes from pushing off the rear leg and rotating the hips and shoulders. To execute the rear hand punch, Marines—

Rotate the hips and shoulders forcefully toward the opponent and thrust the rear hand straight out, rotating the palm down, to nearly full extension.

Shift body weight to the lead foot while pushing off on the ball of the rear foot.

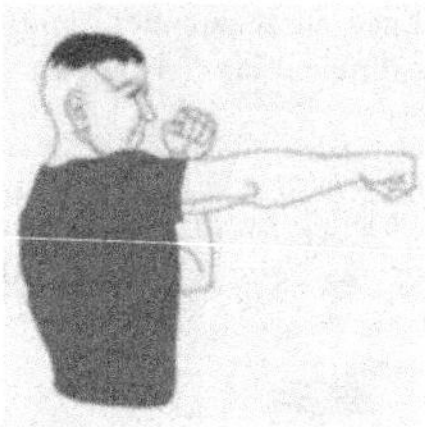

Contact the opponent with the first two knuckles of the fist.

Retract the hand immediately.

Uppercut

The uppercut is a powerful punch originating below the opponent's line of vision. It is executed in an upward motion traveling up the centerline of the opponent's body. It is delivered in close and usually follows a preparatory strike that leaves the target area unprotected. When delivered to the chin or jaw, the uppercut can render an opponent unconscious, cause extensive damage to the neck, or sever the tongue. To execute the uppercut, Marines—

Bend the arms, rotating the palm inboard. The distance the arms bend depends on how close the opponent is.

Rotate the hips and shoulders forcefully toward the opponent, thrusting the fist straight up toward the opponent's chin or jaw.

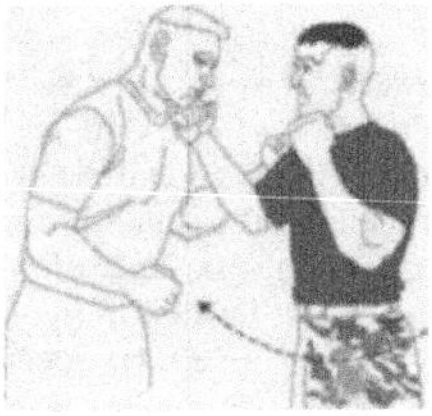

Contact the opponent with the first two knuckles of the fist.

Retract the hand immediately.

Hook

The hook is a powerful punch that is executed close in and is usually preceded by a preparatory strike. To execute the hook, Marines—

Thrust the right arm in a hooking motion toward the opponent, keeping the elbow bent while forcefully rotating the right shoulder and hip toward the opponent.

Contact the opponent with the first two knuckles of the fist. Continue rotating the shoulder and hip, following through with the fist to the target.

Retract the hand immediately.

3. Strikes with the Upper Body

Strikes stun the opponent or set him up for a follow-up finishing technique. The hands, forearms, and elbows are individual weapons of the arms that can be used to execute strikes including the hammer fist, chin jab, knife hand, eye gouge, and elbow strikes. These strikes provide a variety of techniques that can be used in any type of close combat situation.

Principles of Execution

There are several principles of execution that ensure a strike's effectiveness.

Generating Power. Weight transfer is necessary to generate maximum power in a punch. Marines accomplish this by-

- Rotating the hips and shoulders into the attack.
- Moving the body mass straight forward or backward in a straight line.
- Dropping body weight into an opponent. Body mass can be transferred into an attack from high to low or from low to high.

Muscular Tension. The arms are relaxed until the moment of impact. At the point of impact, Marines apply muscular tension in the hand and forearm to maximize damage to the opponent and to avoid injury to the hand. The arms are relaxed until the moment of impact.

Hit and Stick and Follow-Through. A strike should be delivered so that the weapon (e.g., hand, elbow) hits and remains on the impact site (target) and follows through the target. This technique inflicts maximum damage on the opponent.

Strikes with the arms are executed with "heavy hands"; i.e., the strike is executed by driving through with the strike to allow the weight of the hand to go through the target area of the body. Contact with the opponent should be made with the arm slightly bent, and the arm extends as it moves through the target. This technique allows Marines to deliver strikes effectively without executing full force.

Movement. Movement puts Marines in the proper position for launching an attack against an opponent as well as providing protection.

Movement is initiated from the basic warrior stance and ends by resuming the basic warrior stance. Strikes can be performed with either the left or right arm depending upon—

The angle of attack.
The position of the opponent.
The opponent's available, vulnerable target areas.

Target Areas of the Body

For each strike, there are target areas of the body that, when struck, maximize damage to an opponent. Strikes use gross motor skills as opposed to fine motor skills. The target areas of the body are just that—areas. Pinpoint accuracy on a specific nerve is not needed for the strike to be effective.

Hammer Fist

Striking with the hammer fist concentrates power in a small part of the hand, which, when transferred to the target, can have a devastating effect. The striking surface of the hammer fist is the meaty portion of the hand below the little finger. To execute the hammer fist strike, Marines-

Make a fist and bend the arm at approximately a 45- to 90-degree angle. At the same time, rotate the right hip and right shoulder backward.

Thrust the fist forward onto the opponent while rotating the right hip and shoulder forward.

Rotate the wrist so the hammer fist makes contact on the opponent.

Chin Jab

The chin jab can immediately render an opponent unconscious and cause extensive damage to the neck and spine. The striking surface is the heel of the palm of the hand. To execute the chin jab, Marines—

Bend the right wrist back at a 90-degree angle with the palm facing the opponent and the fingers pointing up.

Keep the right arm bent and close to the body. Extend the hand into a concave position with the fingers slightly spread apart.

Step forward with the left foot toward the opponent, keeping the feet approximately shoulder-width apart and the knees bent. This is done to close with the opponent.

Keep the right arm bent and close to the side. Thrust the palm of the hand directly up under the opponent's chin. At the same time, rotate the right hip forward to drive the body weight into the attack to increase the power of the strike. The attack should travel up the centerline of the opponent's chest to his chin.

Knife Hand

The knife hand is one of the most versatile and devastating strikes. The striking surface is the cutting edge of the hand, which is the meaty portion of the hand below the little finger extending to the top of the wrist. The striking surface is narrow, allowing strikes on the neck between the opponent's body armor and helmet. The knife hand strike is executed from one of three angles: outside, inside, and vertical.

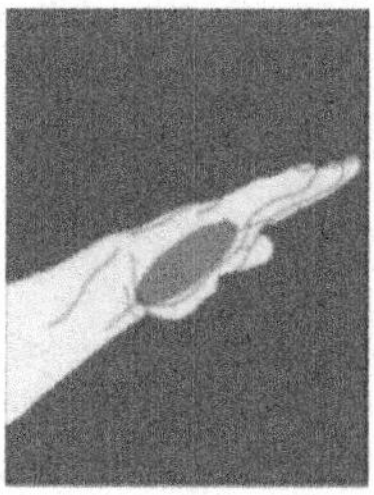

Outside Knife Hand. To execute the outside knife hand strike, Marines—

Execute a knife hand by extending and joining the fingers of the right hand and placing the thumb next to the forefinger (like saluting).

Retract the right hand. At the same time, rotate the right hip and right shoulder backward.

Thrust the knife hand forward (horizontally) onto the opponent while rotating the right hip and shoulder forward.

Inside Knife Hand. To execute the inside knife hand strike, Marines—

Execute a knife hand.

Bring the right hand over the left shoulder. At the same time, rotate the right shoulder forward and the left hip forward.

Thrust the knife hand forward (horizontally) onto the opponent while rotating the right hip and shoulder forward and the left shoulder backward.

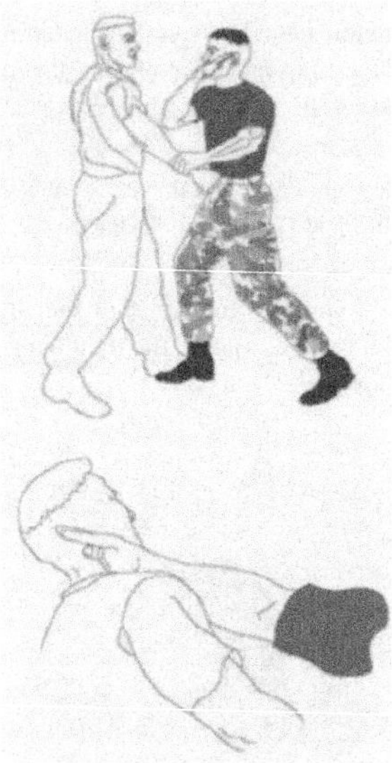

Vertical Knife Hand. When thrown vertically, the knife hand strike comes straight down in a straight line.

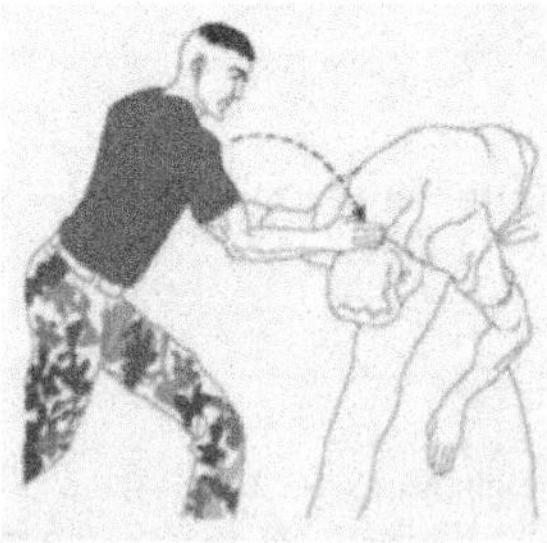

Eye Gouge

The eye gouge is used to attack an opponent's eyes, blinding him so follow-up strikes can be executed. The striking surface is the tips of the fingers or thumb. To execute the eye gouge, Marines—

Extend the right hand with the fingers slightly spread apart to allow entry into the eye sockets.

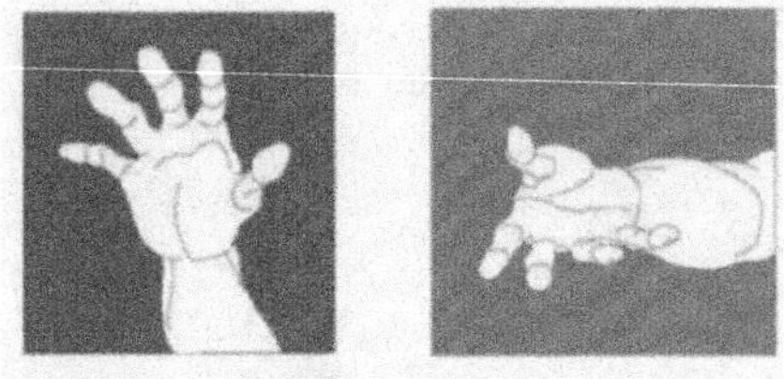

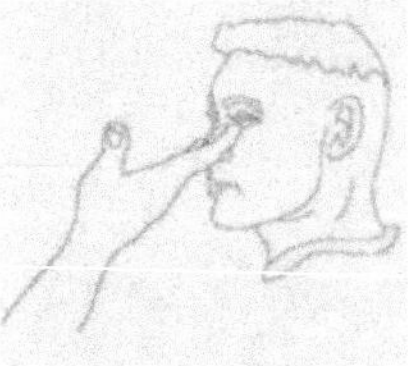

Place the palm of the hand either toward the ground or toward the sky and thrust the right hand forward into the opponent's eyes.

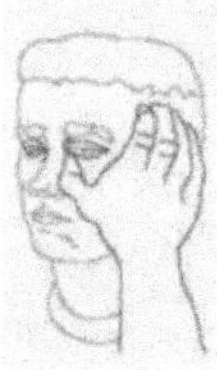

Thrust the hand forward at the opponent's nose level so the fingers or thumb slide naturally into the grooves of the opponent's eye sockets.

Elbow Strikes

The elbow is a powerful weapon that can be used in several different ways to attack virtually any part of an opponent's body. Elbow strikes can be performed either vertically (upward or downward) or horizontally (forward or reverse). The striking surface is 2 inches above or below the point of the elbow, depending upon the angle of attack, the opponent's attack angle, and the position of the opponent.

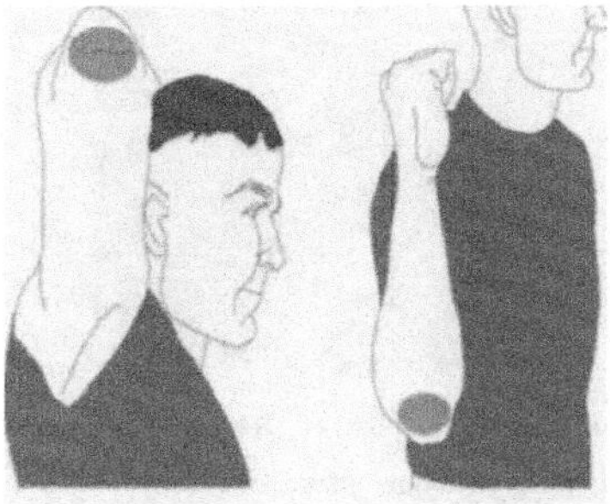

Above Elbow Below Elbow

Vertical Elbow Strike (Up). To execute an upward vertical elbow strike, Marines-

Bend the right elbow, keeping the fist close to the body. The fist is at shoulder level and the elbow is next to the torso.

Thrust the elbow vertically upward toward the opponent while rotating the right shoulder and hip forward to generate additional power.

Contact the opponent with the right forearm 2 inches above the point of the elbow.

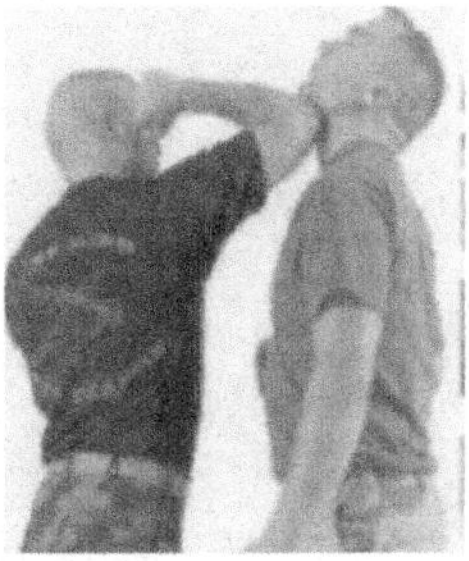

Vertical Elbow Strike (Down). To execute a downward vertical elbow strike, Marines—

Bend the right elbow, keeping the fist close to the body. The fist is on the shoulder and the elbow is raised well above the shoulder.

Thrust the elbow vertically downward toward the opponent while dropping body weight into the attack to generate additional power.

Contact the opponent with the right triceps 2 inches above the point of the elbow.

Horizontal Elbow Strike (Forward). To execute the forward horizontal elbow strike, Marines-

Tuck the right fist near the chest with the palm heel facing the ground.

Thrust the right elbow horizontally forward toward the opponent. The forearm is parallel to the ground.

Rotate the right shoulder and hip forward to generate additional power.

Contact the opponent with the right forearm 2 inches below the point of the elbow.

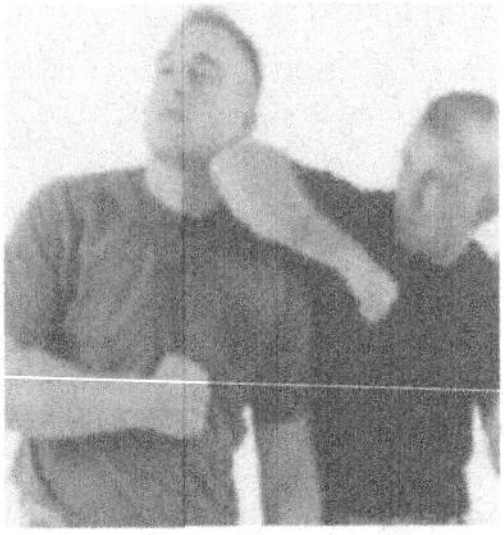

Horizontal Elbow Strike (Rear). To execute the rear horizontal elbow strike, Marines-

Tuck the right fist near the left shoulder with the palm heel facing the ground. At the same time, rotate the right shoulder forward and the left hip forward.

Thrust the right elbow horizontally rearward toward the opponent. The forearm is parallel to the ground and the hand moves toward the direction of the attack.

Rotate the right hip back and the right shoulder backward to generate additional power.

Contact the opponent with the right triceps 2 inches above the point of the elbow.

4. Strikes with the Lower Body

The legs are the body's most powerful weapons because they use the largest muscle groups to generate a strike. Legs are also less prone to injury. The feet are the preferred choice for striking because they are protected by boots. Marines use their feet, heels, and knees to execute kicks, knee strikes, and stomps.

Kicks

The purpose of a kick is to stop an opponent's attack or to create an opening in his defense in order to launch an attack. Kicks can be performed with the left (lead) leg or the right (rear) leg. Kicks with the rear leg have greater power because the hips are rotated into the attack. However, the rear leg is further away from the opponent so a strike with the rear leg will not contact the opponent as quickly as a strike with the lead leg.

Front Kick. The front kick is executed when the opponent is in front of the Marine. The front kick, delivered with either the rear or lead leg, is effective for striking below the waist. Attempting to kick higher results in diminished balance and provides the enemy with a greater opportunity to grab the leg or foot. The striking surfaces arc the toe of the boot or the bootlaces. To execute the front kick, Marines-

Raise the left knee waist high, pivot the hips into the attack, and thrust the left foot forward toward the opponent.

Contact the opponent with the toe of the left boot or bootlaces.

Return to the basic warrior stance.

Side Kick. The side kick, delivered with the lead leg, is effective for striking the knees. The side kick is executed when the opponent is to the side of the Marine. The striking surface is the outside cutting edge of the boot near the heel. To execute the side kick, Marines—

Raise the right knee waist high and rotate the right hip forward.

Thrust the right foot to the right side toward the opponent, turning the foot at a 90-degree angle to maximize the striking surface on the opponent.

Contact the opponent with the cutting edge of the right boot.

Return to the basic warrior stance.

Knee Strikes

Knee strikes are excellent weapons during the close range of close combat fighting. The knee strike is generally delivered in close.

Vertical Knee Strike. The striking surface is the thigh, slightly above the knee. To execute the vertical knee strike, Marines-

Raise the right knee and drive it up forcefully into the opponent. Power is generated by thrusting the leg upward.

Contact the opponent 2 inches above the right knee.

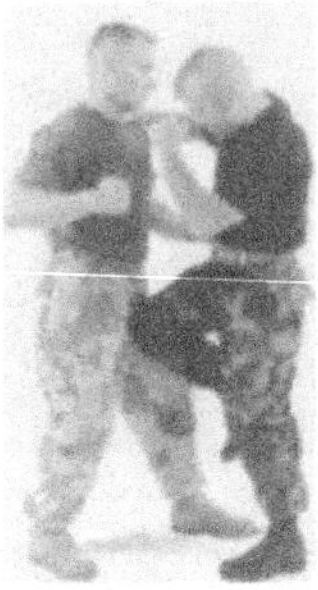

Horizontal Knee Strike. The horizontal knee strike is executed with the leg generally parallel with the ground while rotating the hips to generate power. It is often delivered to the peroneal nerve. The striking surface is the front of the leg, slightly above or below the knee. To execute the horizontal knee strike, Marines-

Raise the right knee, rotate the right hip forward while pivoting on the left foot, and drive the knee horizontally into the opponent.

Contact the opponent 2 inches above the right knee.

Stomps

Stomps are delivered with the feet, usually when the opponent is down. Remember, when the opponent is down, Marines take whatever target is available.

Vertical Stomp. The vertical stomp allows Marines to remain upright and balanced, to rapidly deliver multiple blows with either foot, and to quickly and accurately attack the target. It is the preferred stomp. The striking surface is the flat bottom of the boot or the cutting edge of the heel. To execute the vertical stomp, Marines—

Raise the right knee above the waist with the right leg bent at approximately a 90-degree angle.

Drive the flat bottom of the right boot or the cutting edge of the right heel down onto the opponent forcefully. At the same time, bend the left knee slightly to drop the body weight into the strike.

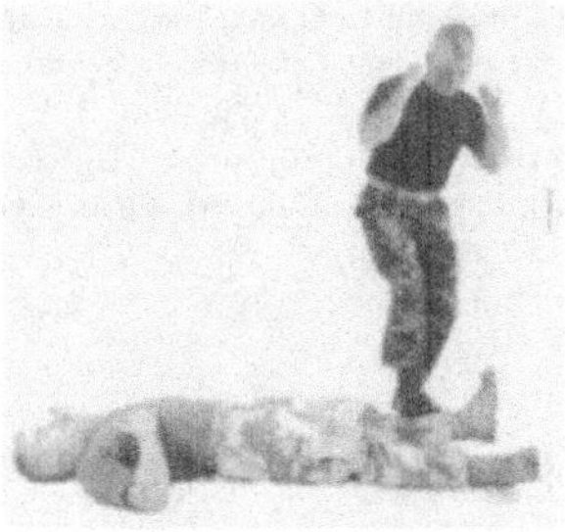

Ax Stomp. The striking surface of the ax stomp is the cutting edge of the heel. To execute the ax stomp, Marines-

Raise the right heel above the waist, keeping the right leg straight.

Drive the cutting edge of the right heel down onto the opponent forcefully. At the same time, bend the left knee slightly to drop the body weight into the strike.

5. Counters to Strikes

In a close combat situation, an opponent will attempt to strike Marines with punches and kicks. When an opponent uses a strike, Marines must first avoid the strike. This is accomplished by moving quickly and blocking. Next, Marines must get into an offensive position. This allows Marines to use offensive strikes to attack the opponent. Regardless of the strike, the counter to a strike requires Marines to move, block, and strike.

Move

The first step in countering a strike is to move out of the way of the strike's impact. Movement removes Marines from the opponent's intended strike point and positions Marines to attack. Movement is executed at approximately a 45-degree angle to the front or rear. Movement is always initiated from the basic warrior stance. Return to the basic warrior stance with the toe of the lead foot pointing toward the opponent once the movement is complete.

Block

Different blocks are executed based on the strike. These will be covered by individual counters.

Strike

Any of the upper body or lower body strikes can be executed as a follow-on attack or part of the counter to an opponent's strike. The follow-on strike is determined by the angle to the opponent, the position of the opponent, and the opponent's available vulnerable target areas.

Counters to Punches

Counter to a Lead Hand Punch. This counter is used when the opponent throws a lead hand punch. To execute the counter to the lead hand punch, Marines-

Step forward and to the left at approximately a 45-degree angle, moving in to the outside of the opponent's attacking arm.

Raise the left arm and block or deflect the opponent's lead hand with the palm of the hand or the meaty portion of the forearm.

Hit and stick by leaving the left arm against the opponent's right arm while stepping forward and to the left at approximately a 45-degree angle to close with the opponent.

Execute a finishing technique, such as a strike or a kick, to the opponent's exposed target areas.

Counter to a Rear Hand Punch. This counter is used when the opponent throws a rear hand punch. To execute the counter to the rear hand punch, Marines—

Step forward and to the left at approximately a 45-degree angle, moving in to the outside of the opponent's attacking arm.

Raise the left arm and block or deflect the opponent' s rear hand with the palm of the hand or the meaty portion of the forearm.

Hit and stick by leaving the left arm against the opponent's right arm while stepping forward and to the right at approximately a 45-degree angle to close with the opponent.

Execute a finishing technique, such as a strike or a kick, to the opponent's exposed target areas.

Counters to Kicks

Counter to a Front Kick (Left or Lead Leg). This counter is used when the opponent executes a front kick with his left leg. To execute the counter to a front kick, Marines-

Step forward and to the right at approximately a 45-degree angle, moving in to the outside of the opponent's striking leg.

Lower the left arm and block or deflect the opponent's leg with the palm of the hand or the meaty portion of the forearm.

Hit and stick by leaving the left arm against the opponent's leg while stepping forward and to the left at approximately a 45-degree angle to close with the opponent.

Execute a finishing technique, such as a strike or a kick, to the opponent's exposed target areas.

Counter to a Front Kick (Right or Rear Leg). This counter is used when the opponent executes a front kick with his right leg. To execute the counter to a front kick, Marines—

Step forward and to the left at approximately a 45-degree angle, moving in to the outside of the opponent's striking leg.

Lower the left arm and block or deflect the opponent's leg with the palm of the hand or the meaty portion of the forearm.

Hit and stick by leaving the left arm against the opponent's leg while stepping forward and to the right at approximately a 45-degree angle to close with the opponent.

CHAPTER 5

THROWS

This chapter describes all techniques for a right-handed person. However, all techniques can be executed from either side.

The Marine is depicted in camouflage utilities. The opponent is depicted without camouflage.

Marines use throwing techniques to maintain the tactical advantage and to throw the opponent to the ground during close combat. Throws apply the principles of balance, leverage, timing, and body position to upset an opponent's balance and to gain control by forcing the opponent to the ground. Throwing techniques are effective because they are size- and gender-neutral, and they rely on the momentum and power generated by the opponent rather than the strength or size of the Marine. Marines also execute a throw as a devastating attack against an opponent, possibly causing unconsciousness or broken limbs. When Marines execute throws, they must maintain balance and, simultaneously, prevent the opponent from countering a throw or escaping after being forced to the ground.

1. Turning Throw

Marines use a turning throw to take the opponent to the ground while they remain standing. A turning throw can also be executed from a stationary position. It is particularly effective if Marines and the opponent are wearing gear. To execute the turning throw, Marines-

Grasp the opponent's right wrist with the left hand.

Step forward with the right foot, place it against the outside of the opponent's right foot, and pivot so the back of the heel is next to the middle of the opponent's foot.

Hook the opponent's right arm with the right arm and pinch his arm between the biceps and forearm, touching the opponent with the body.

Pull the opponent's wrist downward, keeping it close to the body.

Pivot to the left on the ball of the foot and continue pulling downward on the opponent's wrist while rotating the wrist outward to off-balance him.

2. Hip Throw

Marines use a hip throw to take the opponent to the ground while they remain standing. A hip throw is particularly effective if the opponent is moving forward or pushing. Marines use the opponent' s forward momentum to execute the hip throw. To execute the hip throw, Marines-

Grasp the opponent's right wrist with the left hand.

Step forward with the right foot and place it against the outside of the opponent's right foot.

The back of the heel should be next to the opponent' s foot.

Step back with the left foot and rotate on the ball of the foot. The back of the heel is next to the opponent' s toe. The knees are bent.

Rotate at the waist and hook the right hand around the back of the opponent's body (anywhere from his waist to his head). The side and hip should be against the opponent.

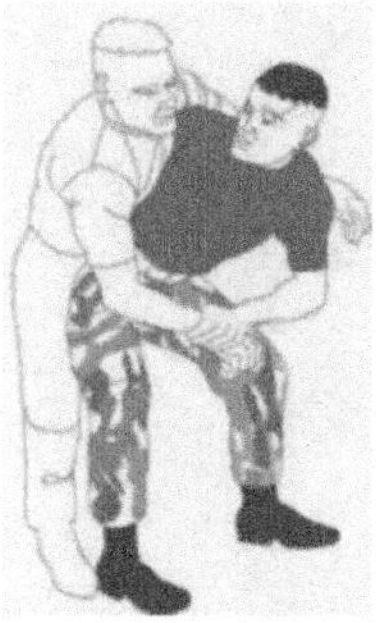

Rotate the hip against the opponent. The hips must be lower than the opponent's.

Use the right hand to pull the opponent up on the hip to maximize contact.

Pull the opponent's arm across the body and, at the same time, lift the opponent off the ground slightly by bending at the waist, straightening the legs, and rotating the body to the left.

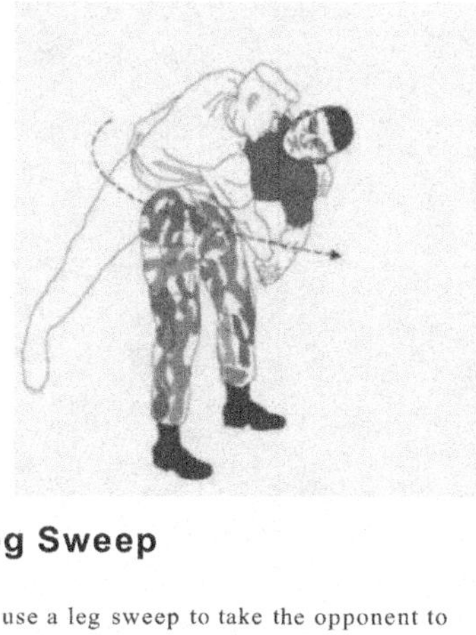

3. Leg Sweep

Marines use a leg sweep to take the opponent to the ground while they remain standing. A leg sweep is particularly effective if the opponent is already off-balanced and moving backward or pulling on the Marine. To execute the leg sweep, Marines—

Grab the opponent's right wrist with the left hand and grab the opponent' s left shoulder with the right hand.

Note: Grab the opponent's clothing or gear if his wrist and shoulder cannot be grabbed.

Step forward with the left foot and place it on the outside of the opponent's right foot.

Pull the opponent's wrist downward, close to the body, and push his shoulder backward to off-balance him.

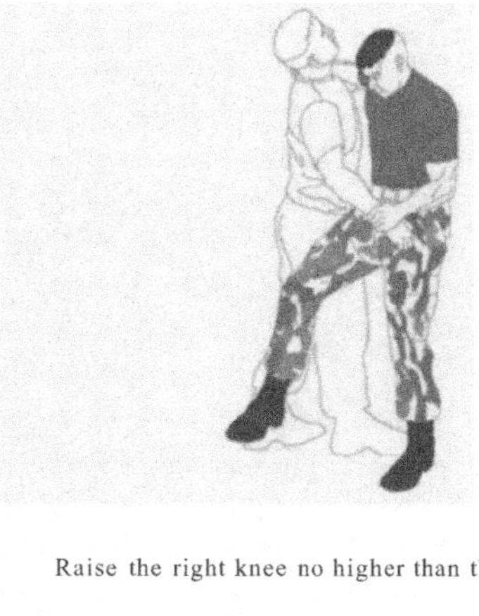

Raise the right knee no higher than the waist.

Kick the foot past the opponent's right leg.

Use the heel of the boot to make contact with the opponent's calf (anywhere from the top of the calf down to the Achilles tendon or on the inside of the calf).

Sweep the opponent to the ground.

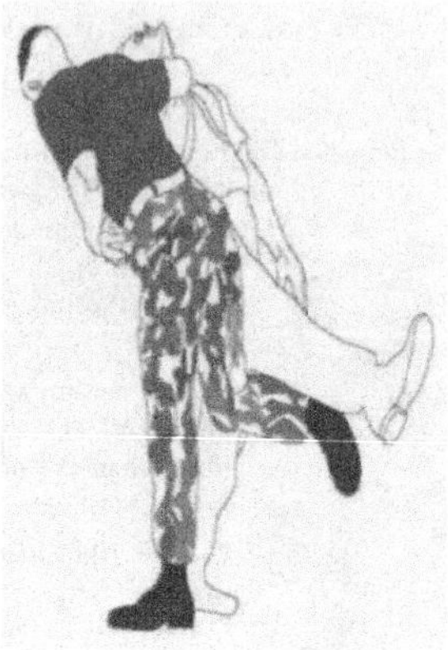

CHAPTER 6

CHOKES AND HOLDS

This chapter describes all techniques for a right-handed person. However, all techniques can be executed from either side.

The Marine is depicted in camouflage utilities. The opponent is depicted without camouflage.

WARNING

During training, never execute a choke at full force or full speed and never hold a choke for more than 5 seconds.

When Marines correctly perform a choke, they render an opponent unconscious in as little as 8 to 13 seconds. Chokes are easily performed regardless of size or gender. Marines must know how to apply chokes and how to counter a choke or a hold executed by an opponent.

1. Types of Chokes

There are two types of chokes: an air choke and a blood choke. An air choke closes off the airway to the lungs, thereby preventing oxygen from reaching the heart. A blood choke cuts off the blood flow to the brain. Both types can result in unconsciousness and eventual death for an opponent.

Air Choke

An air choke is performed on the opponent's windpipe (or trachea), cutting off the air to the lungs and heart. If Marines execute the air choke properly, the opponent loses consciousness within 2 to 3 minutes. Due to the length of time it takes to immobilize the opponent, air chokes are not recommended.

Blood Choke

A blood choke is performed on the opponent's carotid artery, which carries oxygen-enriched blood from the heart to the brain. The carotid artery is located on both sides of the neck. If Marines execute a blood choke properly, the opponent will lose consciousness within 8 to 13 seconds. The blood choke is the preferred choke because its intended effect (i.e., the opponent losing consciousness) can be executed quickly, ending the fight.

2. Chokes

Front Choke

Marines execute a front choke when they are facing the opponent. Marines use the opponent's lapels or collar to execute a front choke. To execute the front choke, Marines-

Grab the opponent's right lapel with the right hand, making certain that the knuckles or the back of the hand are against the opponent's carotid artery on the right side of his neck.

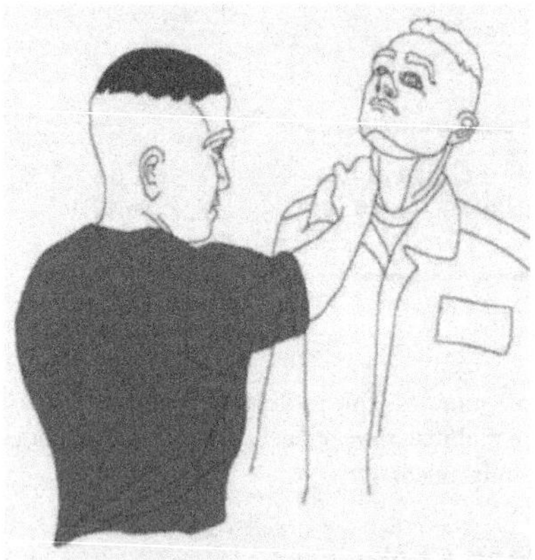

Keep the right hand against the opponent's neck, reach under the opponent's right arm with the left hand, grab the opponent's left lapel, and form an X with the wrists.

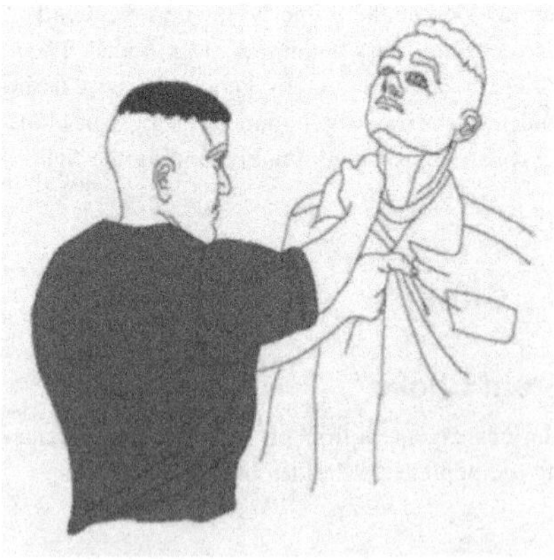

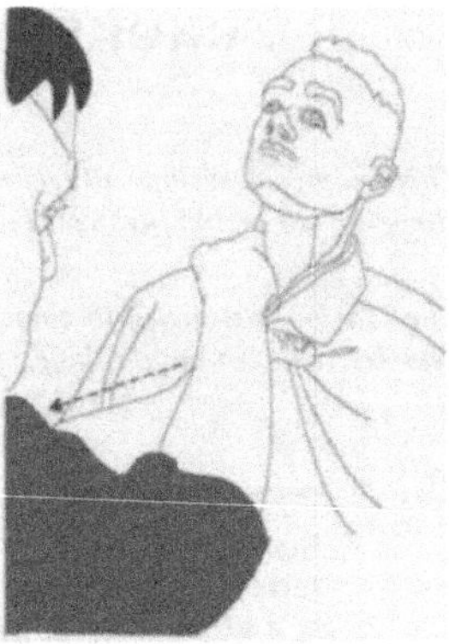

Press the right hand against the opponent's carotid artery.

Pull his lapel to the left with the left hand.

Side Choke

Marines execute a side choke when they are facing the opponent. The side choke is particularly effective when deflecting a punch thrown by an opponent. To execute the side choke, Marines—

Use the left hand to parry the opponent's arm inboard (to the inside of the opponent's reach).

Bring the right arm underneath the opponent's arm and up around the front of his neck.

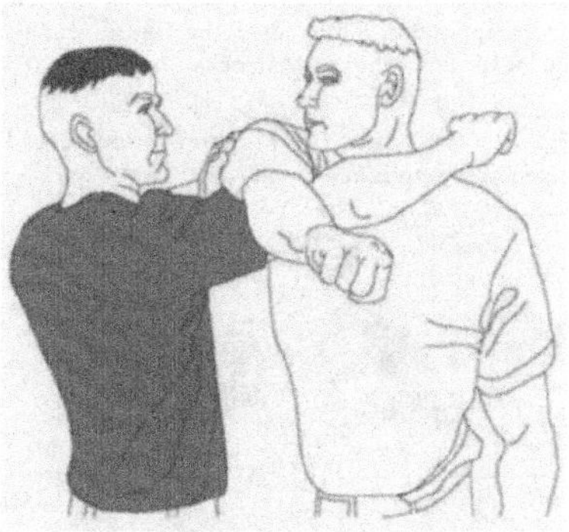

Extend the fingers, place the back of the forearm against the opponent's neck just below his ear, and press the carotid artery on his neck.

Note: If Marines are unable to place their fingers against the opponent's neck, they may use the back of the thumb or wrist.

Reach, with the left hand, around the back of the opponent's neck and clasp the hands together.

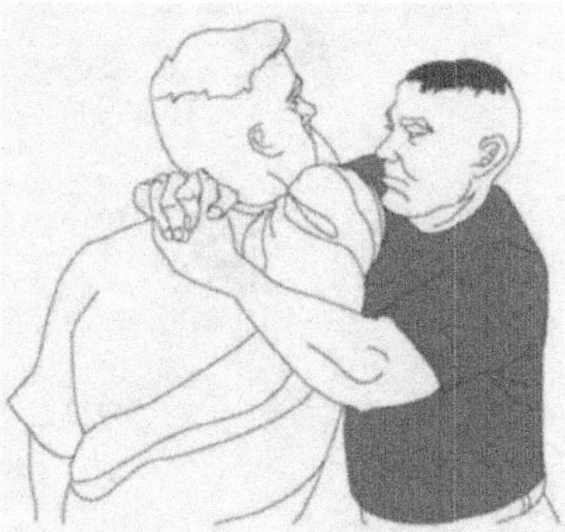

Pull the opponent toward the chest by pulling the clasped hands toward the chest.

Exert pressure on the side of the opponent's neck with the forearm.

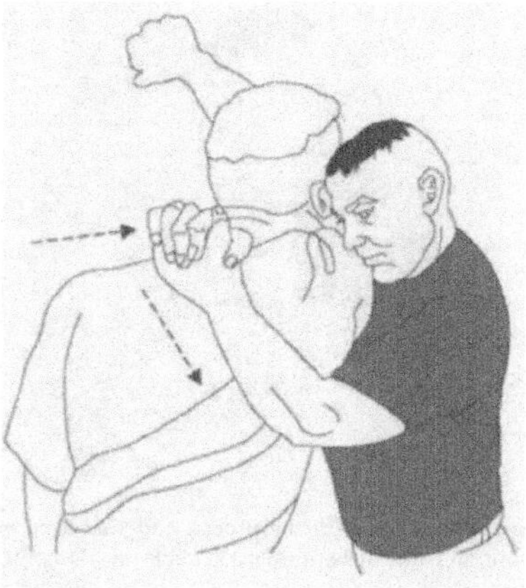

Rear Choke

Marines execute a rear choke when they are behind the opponent, the opponent is on the ground, or when they are taking the opponent to the ground. To execute the rear choke, Marines—

Reach, with the right arm, over the opponent's right shoulder and hook the bend of the arm around his neck.

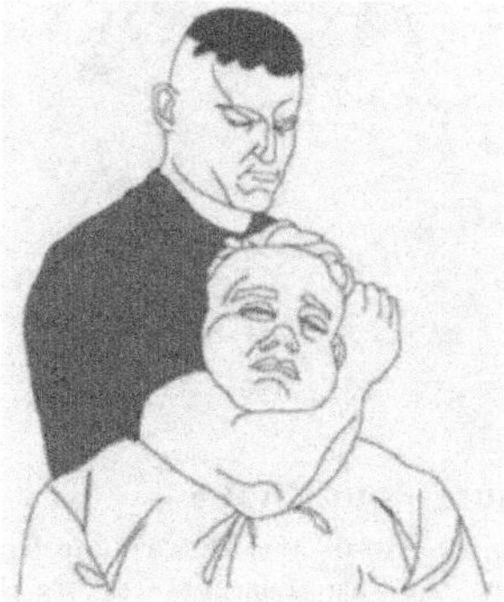

Clasp both hands together.

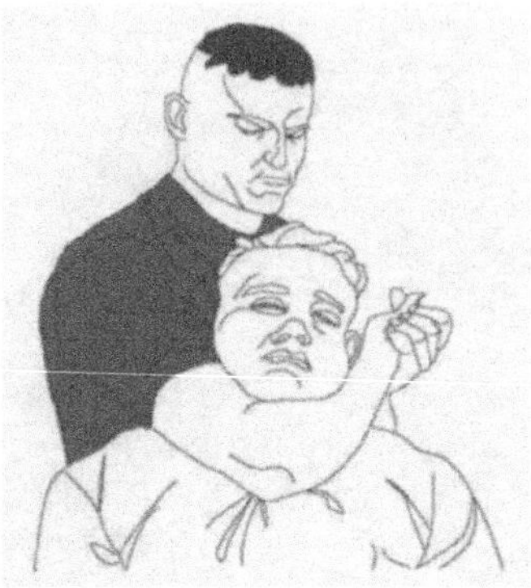

Exert pressure with the biceps and forearms on both sides of the opponent's neck on his carotid arteries.

Maintain pressure with the biceps and forearms on both sides of the neck and draw the opponent closer by drawing the right arm in.

Figure-Four Choke

A variation of the rear choke is the figure-four choke. The figure-four choke allows Marines to gain more leverage than the rear choke. If the rear choke cannot be secured, the figure-four variation is employed to increase the pressure of the choke on the opponent. To execute the figure-four variation of the rear choke, Marines—

Apply a rear choke. The Marine's body should be against the opponent's body.

Grasp the left biceps with the right hand and place the left hand against the back of the opponent's head.

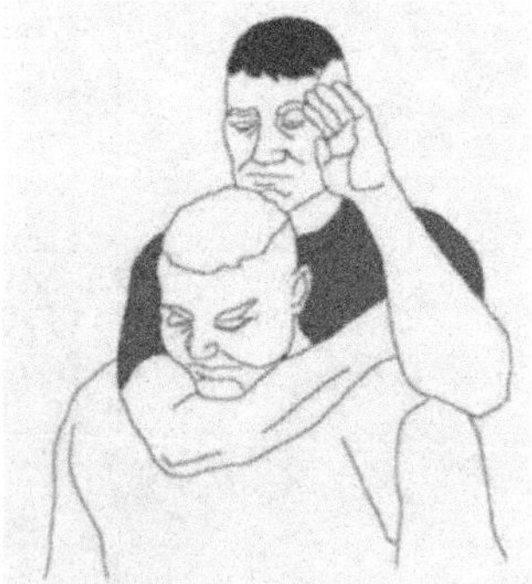

Push the opponent's head forward and down with the left hand.

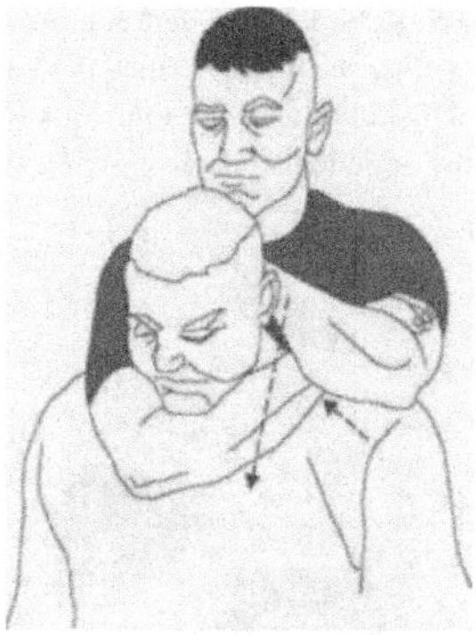

Draw the right arm in, maintaining pressure with the bicep and forearm on both sides of the opponent's neck.

3. Counters to Chokes and Holds

During a close combat situation, an opponent may apply a choke or hold on a Marine. If the opponent correctly applies a choke, a Marine quickly loses consciousness. If a choke is not executed properly, it often results in a hold, typically a bear hug or a headlock. A hold allows the opponent to control a Marine and removes the Marine's ability to attack. It is important for Marines to extract themselves from chokes and holds, regain the tactical advantage, and counter with strikes.

Although a choke causes unconsciousness in 8 to 13 seconds for a blood choke and 2 to 3 minutes for an air choke, the first movement in any counter to a choke is to clear the airway. Marines use softening techniques to loosen an opponent's grip and to clear their airway. Softening techniques are particularly effective if Marines lack the physical strength of their opponent. These techniques include groin strikes, eye gouges, foot stomps, etc. Softening techniques are not offensive; rather, they are used to loosen an opponent's hold.

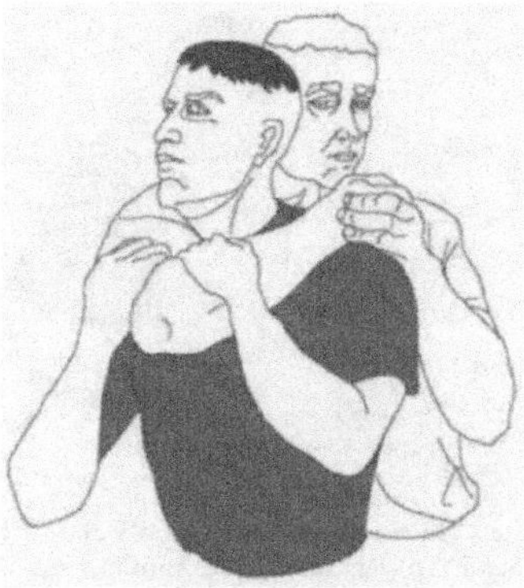

The second movement is to tuck the chin. Once the airway is clear, Marines tuck their chins to prevent the opponent from reapplying the choke.

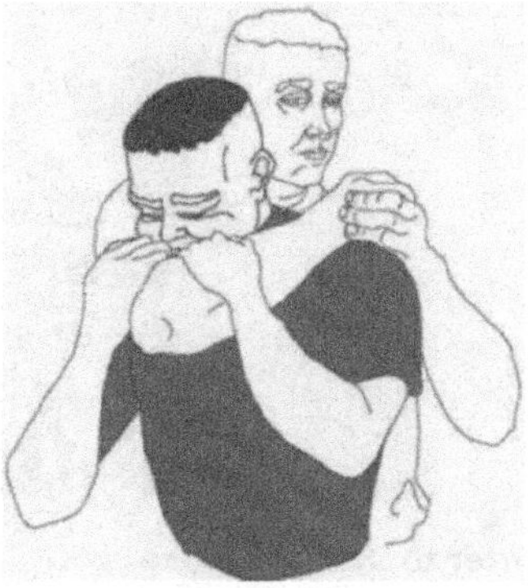

Counter to a Front Choke

Marines use a counter to a front choke when the opponent approaches from the front and uses both hands to choke a Marine around the throat. To execute the counter to the front choke, Marines—

Grasp the opponent's right forearm (where the elbow bends) with the left hand and apply downward pressure on the opponent's radial nerve with the fingers.

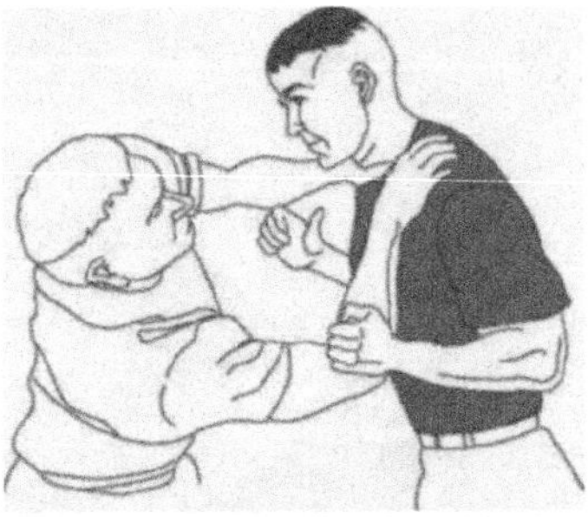

Execute a chin jab to the opponent's chin with the right hand. To generate power into the strike, bring the left foot to the outside of the opponent's right foot and rotate the hips into the strike.

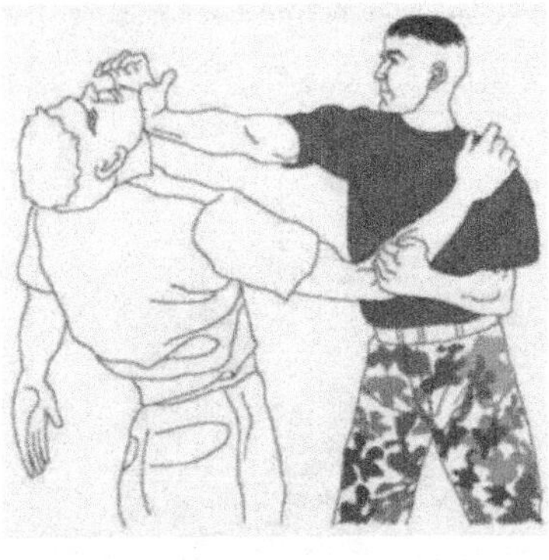

Counter to a Rear Choke

Marines execute a counter to a rear choke when the opponent approaches from the rear and puts his right arm around a Marine's throat. To execute the counter to the rear choke, Marines-

Grasp the opponent's forearm (at the radial nerve) and bicep with both hands and pull down just enough to clear the airway. Once the airway is clear, tuck the chin to protect the airway and to prevent the opponent from reapplying the choke.

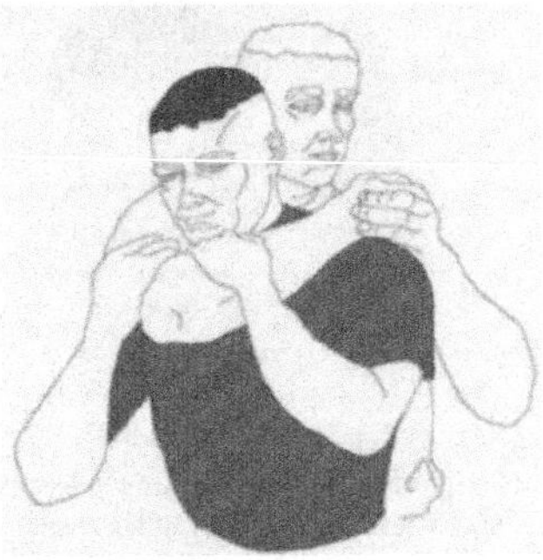

Step behind the opponent's right leg with the left foot, keeping both legs bent (almost in a squatting position).

Strike and drive the left elbow into the opponent's torso while rotating the hips and pivoting to the left, throwing the opponent back and to the ground.

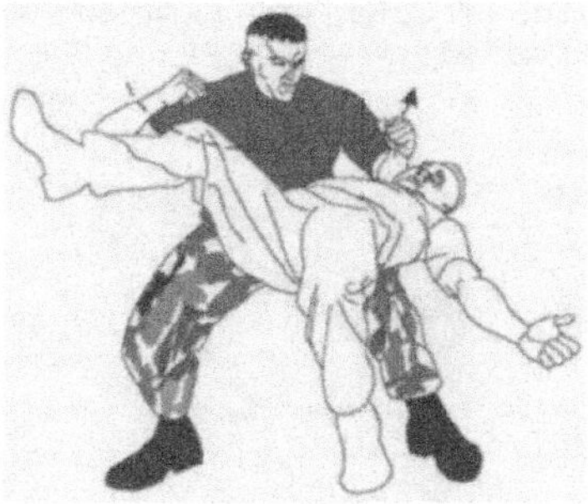

Counter to a Front Headlock

Marines use a counter to a front headlock when the opponent approaches from the front and puts his right arm around the Marine's neck, bends the Marine forward, and locks the Marine's head against his hip. To execute the counter to a front headlock, Marines—

Grasp the opponent's wrist and forearm with both hands and pull down to clear the airway. Maintain control of the opponent's wrist throughout the move. Once the airway is clear, tuck the chin to protect the airway and to prevent the opponent from reapplying the choke.

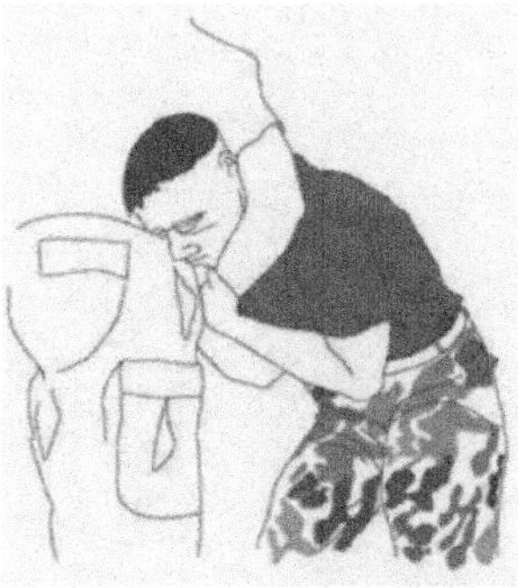

Move the right hand and arm across the opponent's torso.

Step forward and to the left with the left foot at a 45-degree angle.

Execute a sweep with the right foot against the opponent's right leg. At the same time, push against the opponent's chest with the right arm and shoulder to generate power in the sweep.

Counter to a Rear Headlock

Marines use a counter to a rear headlock when the opponent approaches from the rear and puts his right arm around the Marine's neck, bends the Marine forward, and locks the Marine's head against his hip. To execute the counter to the rear headlock, Marines-

Grasp the opponent's wrist and forearm with the right hand and pull down to clear the airway. Once the airway is clear, tuck the chin to protect the airway and to prevent the opponent from re-applying the choke.

Reach over the opponent's right shoulder with the left arm.

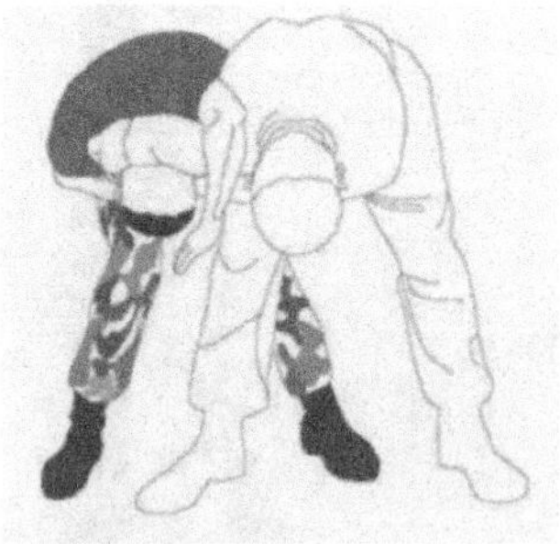

Grab any part of the opponent's face (chin, nose, eyes) and pull back while rising to a standing position.

Execute, with the right hand, a hammer fist strike to the opponent's exposed throat.

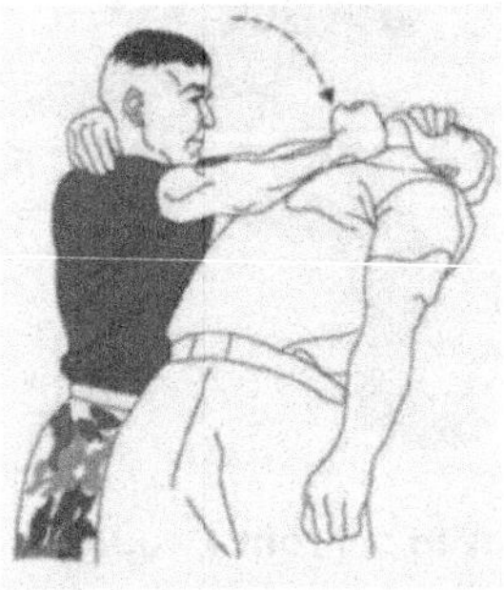

Counter to a Front Bear Hug

Marines execute a counter to a front bear hug when the opponent approaches from the front and puts both of his arms around the Marine's body, trapping the Marine's arms to the sides. To execute the counter to a front bear hug, Marines—

Step forward and to the left with the left foot at a 45-degree angle to the outside of the opponent's right leg, keeping the left leg bent.

Grasp the opponent's torso or arms to gain balance and to assist in throwing the opponent. It may be helpful to hook the opponent's right arm with the left arm.

Drive the right arm and shoulder forward and, at the same time, bring the right leg forward and sweep the opponent's right leg, bringing him to the ground.

Counter to a Rear Bear Hug

Marines use a counter to a rear bear hug when the opponent approaches from the rear and puts both of his arms around the Marine's body, trapping the Marine's arms to the sides. To execute the counter to a rear bear hug, Marines—

Step behind the opponent's right leg with the left foot, keeping both legs bent (almost in a squatting position). The left side of the body should be against the opponent's.

Pivot the hip, turning the body to the left and throwing the opponent back over the bent leg.

CHAPTER 7

GROUND FIGHTING

This chapter describes all techniques for a right-handed person. However, all techniques can be executed from either side.

In drawings, the Marine is depicted in woodland camouflage utilities; the opponent is depicted witlwut camouflage. In photographs, the Marine is depicted in woodland camouflage utilities; the opponent is depicted in desert camouflage utilities.

Marines should avoid being on the ground during a close combat situation because the battlefield may be covered with debris and there is an increased risk of injury. However, many close combat situations involve fighting on the ground. The priority in a ground fight is for Marines to get back on their feet as quickly as possible. In any ground fighting scenario, Marines will usually end up in one of four positions with the opponent. The offensive positions, in which Marines have a tactical advantage, are the guard and mount. The defensive positions, which are used as counters when the opponent has the tactical advantage, are the counter to the guard and the counter to the mount. Marines can also employ chokes during ground fighting to quickly end a fight.

1. Offensive Ground Fighting

Guard Position

Marines execute a guard position when the opponent is on top and the legs are wrapped around the opponent's legs. If the opponent is on top, he may try to choke the Marine, but the Marine still has the tactical advantage because the Marine is in a position to control the situation. To execute the guard position technique, Marines—

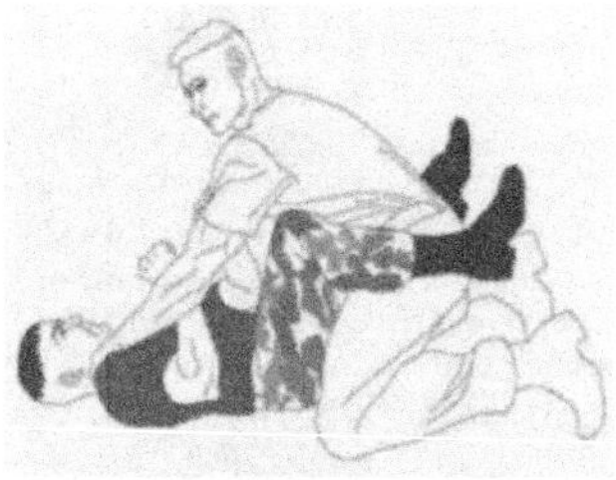

Trap the opponent's hands on the chest by crossing the hands on the chest so the forearms are resting on the opponent's forearms. Apply pressure with the elbows or forearms.

Strike the outside of the opponent's thigh with the cutting edge of the right heel. This causes the opponent to jerk to that side.

Move the head quickly to the left and swivel the hips to the right. At the same time, bring up both of the legs. Both legs are on the right side of the opponent's body.

Bring the right leg down, hooking the opponent's neck and head. Exert downward pressure to roll him over on his back. Grasp and maintain control of the opponent's left arm. Upon completion of the move, the Marine is sitting up with legs bent over the opponent while maintaining control of his arm.

Keep legs and knees bent. Maintain pressure against the opponent's neck with the back of the right foot and against his side with the left foot underneath his armpit. Squeeze the knees together, locking the opponent' s arm.

Pull the opponent's arm straight up and fall back sharply, pulling his arm to the side in the direction of his little finger.

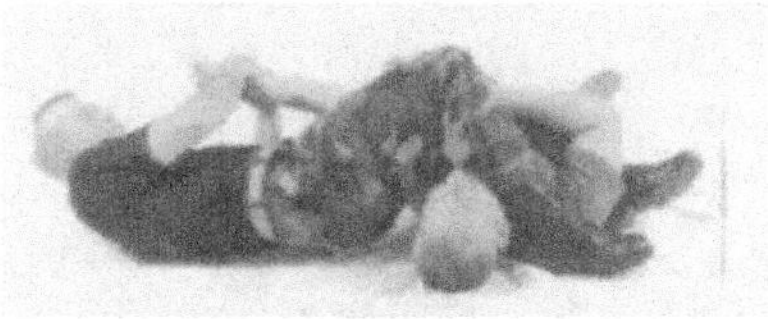

Return to standing.

Mount Position

Marines execute the mount if the opponent is lying on his back on the ground and the Marine is on top with legs wrapped around the opponent's body. This position is an offensive position because Marines are in a better position to control the opponent and to execute ground fighting techniques. To execute the mount position technique, Marines—

Grab the opponent's wrists or forearms with the hands. Hold them tightly against the torso.

Maintain control of the opponent's arms with the left arm and apply pressure to the opponent's brachial plexus (tie in) with the right hand.

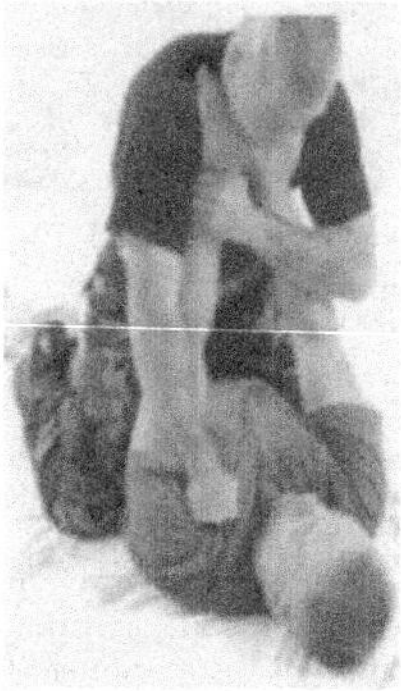

Bring the left foot close to the opponent's right armpit and swing the right leg across the opponent' s head.

Keep the legs and knees bent. Maintain pressure against the opponent's neck with the back of the right foot and against his side with the left foot which is underneath his armpit. Squeeze the knees together, locking the opponent's arm.

Pull the opponent's arm straight up and fall back sharply, pulling his arm to the side in the direction of his little finger.

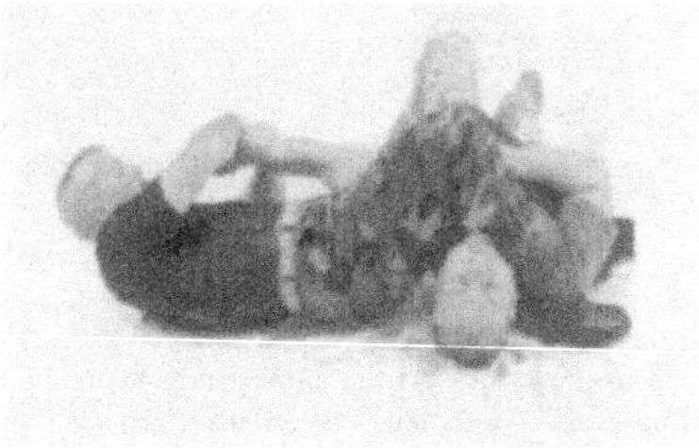

Return to standing.

2. Defensive Ground Fighting

Counter to the Guard

Marines use this technique if the opponent is lying on his back on the ground and the Marine is kneeling on the ground between the opponent's legs. To execute the counter to the guard position, Marines—

Strike the opponent's femoral nerve, located on the inside of the thigh, with the elbows. This forces the opponent to separate his legs.

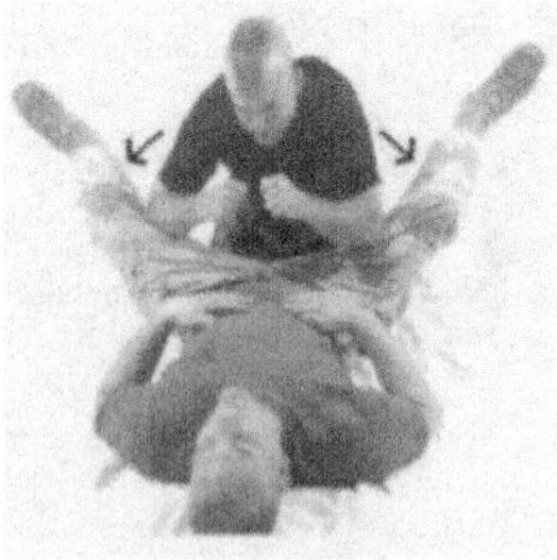

Strike the opponent's groin with the right fist.

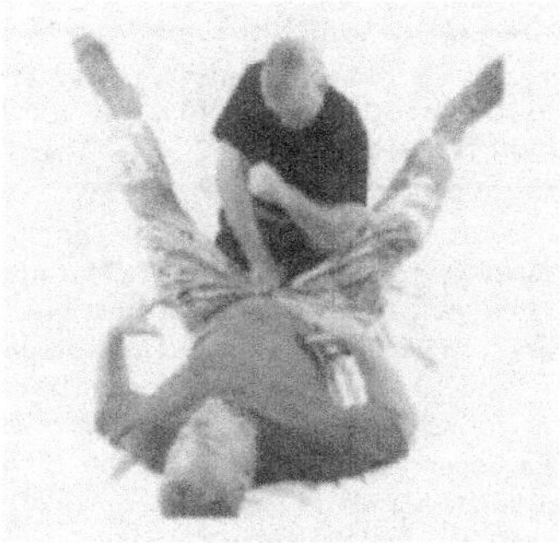

Hook the left arm underneath the opponent's right knee from the inside. Throw the opponent's leg over the head.

Duck the head quickly and move to the left.

Return to standing.

Counter to the Mount

Marines use this technique when they are lying on their back on the ground, the opponent is mounted on top of the Marine, and the opponent's legs are wrapped around the Marine. The opponent has the tactical advantage. To execute the counter to the mount position, Marines—

Grab the opponent's gear or clothing on his upper torso and pull him down close.

Use the right arm to hook the opponent's left arm, from the inside around the outside, above his elbow. With the right foot, hook the opponent's left leg or ankle.

Draw the elbow in to bend the opponent's elbow, bringing him down close. The arm must be hooked above the opponent's elbow in order to bend it.

Strike the opponent's side with the left hand.

Push the opponent over and roll him off to the right side.

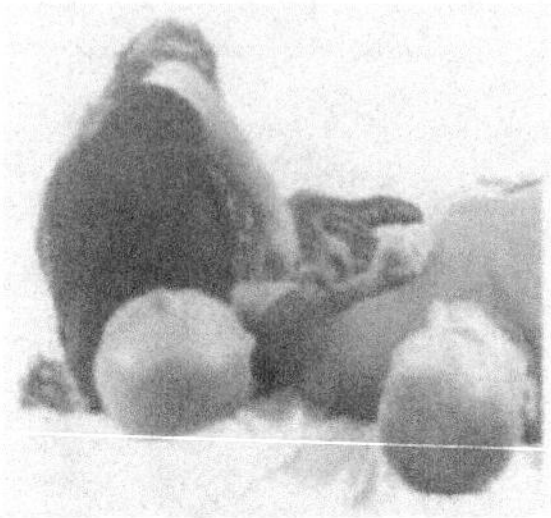

Maintain control of the opponent's hooked arm and move to a kneeling position.

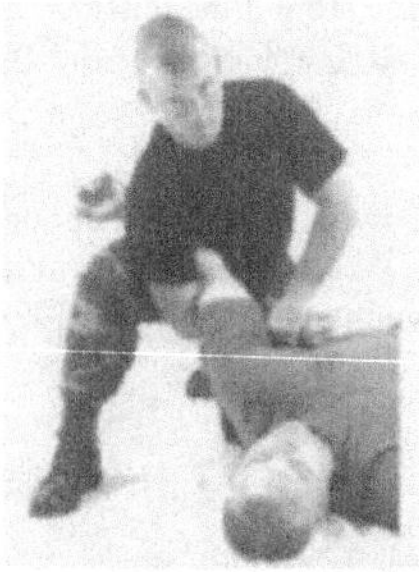

Move to a standing position while maintaining pressure on the opponent's arm and using the knee to apply pressure against the opponent's elbow.

3. Ground Fighting Chokes

WARNING

During training, never execute a choke at full force or full speed. Never hold a choke for more than 5 seconds.

The priority in ground fighting is for Marines to get back on their feet as quickly as possible. Sometimes, Marines can quickly end a ground fight by executing a choke on the opponent. When performed correctly, a choke can render an opponent unconscious in as little as 8 to 13 seconds. Chokes are easily performed regardless of size or gender. The chokes performed during ground fighting are the same as those performed while standing (see chap. 6).

Ground Fighting Front Choke

The ground fighting front choke is a blood choke performed most effectively from the mount position. The front choke employs the opponent's lapels or collar to execute the choke. To execute the ground fighting front choke, Marines—

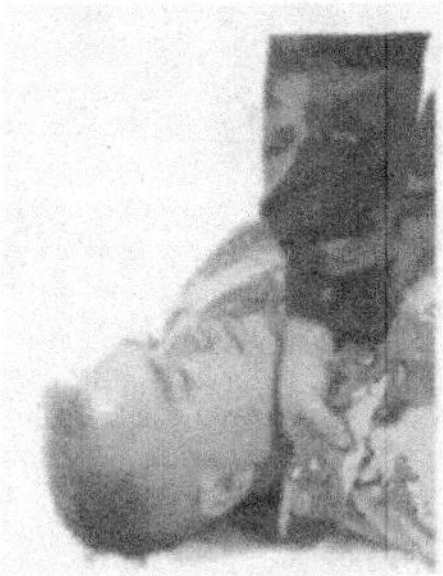

Grab the opponent's right lapel with the right hand, making certain that the knuckles or the back of the hand are against the carotid artery on the right side of the opponent's neck.

Keep the right hand pressed against the opponent's carotid artery, reach under the right arm with the left hand, and grab the opponent's left lapel, forming an X with the wrists.

Keep the right hand pressed against the opponent's carotid artery and pull the opponent's left lapel to the left with the left hand.

Ground Fighting Side Choke

The ground fighting side choke is a blood choke performed from the mount position. The ground fighting side choke is particularly effective when the opponent raises his arms and places them on the Marine's chest or throat. To execute the ground fighting side choke, Marines-

Use the left hand to parry the opponent's right arm inboard (to the inside of the opponent's reach). Bring the right arm underneath the opponent's arm and up around the front of his neck.

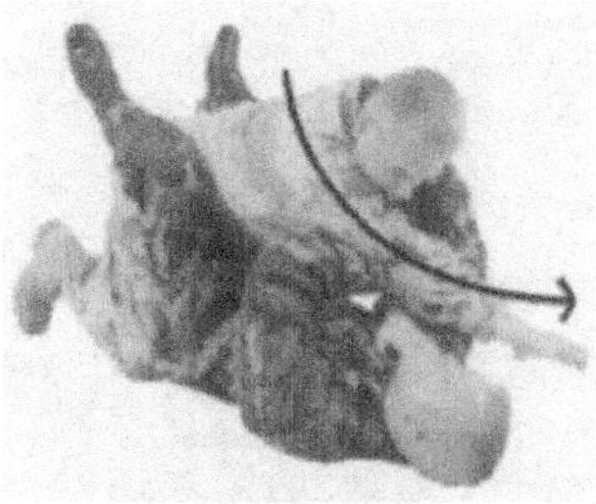

Keep the fingers extended, place the back of the forearm against the opponent's neck just below his ear and press the carotid artery on his neck. Depending on the Marine's or the opponent's body configuration, the back of the thumb or wrist may press against the opponent's neck.

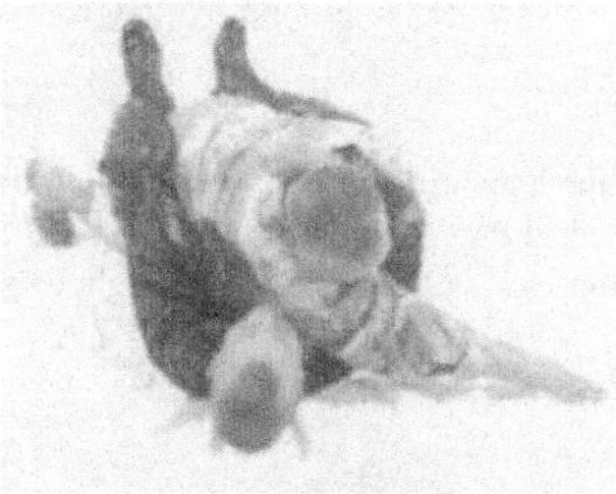

Use the left hand to reach around the back of the opponent's neck and clasp the hands together.

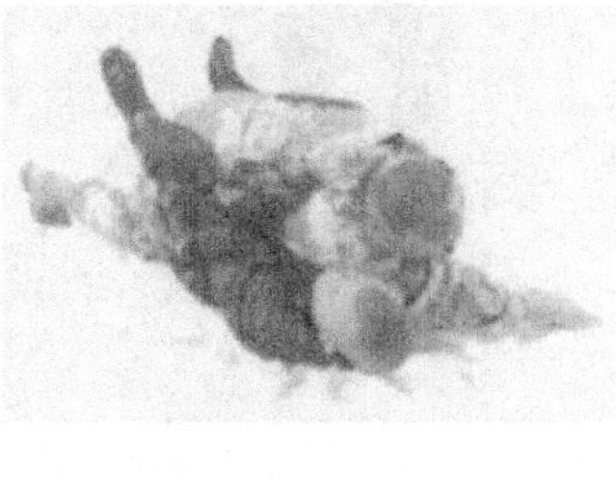

Pull the opponent toward the chest. Use the fore-arm to exert pressure on the side of his neck. This is done by pulling the clasped hands toward the chest.

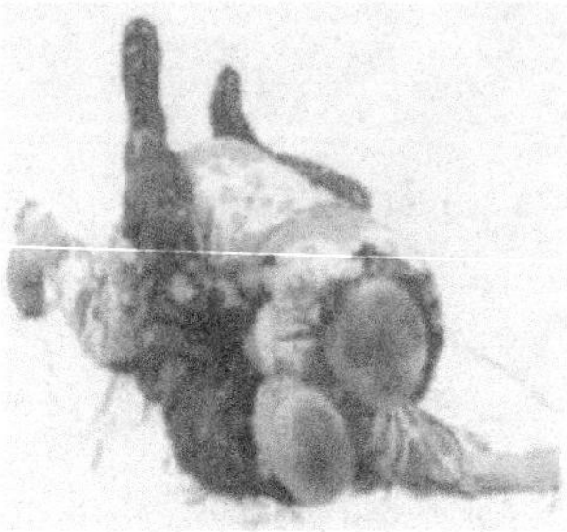

Ground Fighting Rear Choke

The ground fighting rear choke is a blood choke performed when Marines are behind the opponent. To execute the ground fighting rear choke, Marines—

Wrap the legs around the opponent with the left foot against the inside of his left thigh and the right foot against the inside of his right thigh.

Use the right arm to reach over the opponent's right shoulder and hook the bend of the arm around his neck.

Use the left hand to clasp both hands together.

Exert pressure with the biceps and forearms on both sides of the opponent's neck. This applies pressure to the opponent's carotid arteries. While maintaining pressure with the biceps and fore-arms, draw the opponent closer by drawing the right arm in. To increase the effectiveness of the choke, lean backward by arching the back and pulling the opponent back. At the same time, push the feet against the opponent's thighs.

Figure-Four Choke

A variation of the ground fighting rear choke is the figure-four choke. The figure-four choke allows Marines to gain more leverage on the rear choke. If the rear choke cannot be secured, the figure-four variation may be applied to increase the pressure of the choke on the opponent. To execute the figure-four variation of the rear choke, Marines—

Apply the ground fighting rear choke. The chest should be against the opponent's back.

Grasp the left biceps with the right hand and place the left hand against the back of the opponent's head.

Use the left hand to push the opponent's head forward and down.

Draw the right arm in while maintaining pressure with the biceps and forearm on both sides of the opponent's neck.

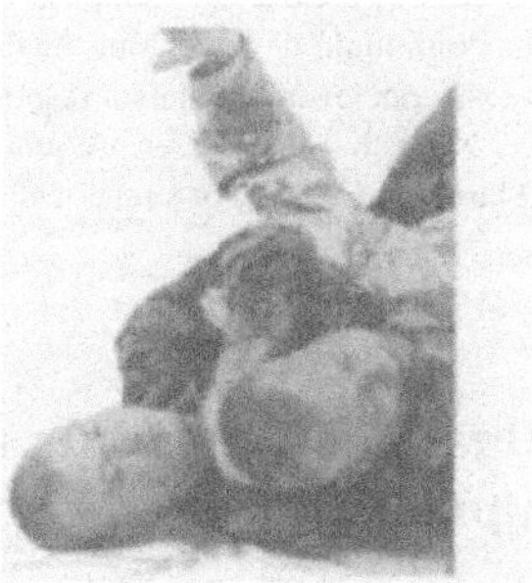

Increase the effectiveness of the choke by arching backward and pushing the feet against the opponent's thighs. Continue to pull the opponent back with the right arm while exerting pressure forward with the left hand against the opponent's head.

CHAPTER 8

NONLETHAL TECHNIQUES

This chapter describes all techniques for a riglit-handed person. However, all techniques can be executed from either side.

In drawings, the Marine is depicted in woodland camouflage utilities; the opponent is depicted without camouflage. In photographs, the Marine is depicted in woodland camouflage utilities; the opponent is depicted in desert camouflage utilities.

The Marine Corps' involvement in military operations other than war—e.g., humanitarian, peacekeeping, or evacuation missions—has greatly increased. These missions require skills that span the spectrum of conflict and support operations within a continuum of force. But the Corps' day-to-day existence also demands a responsible use of force. Nonlethal techniques are among the skills Marines use to apply a responsible use of force.

1. Unarmed Restraints and Manipulation

Marines operate within a continuum of force, particularly in support of peacekeeping or humanitarian types of missions. In these situations, Marines must act responsibly to handle a situation without resorting to deadly force. Unarmed restraints and manipulation techniques including joint manipulation, come-alongs, and takedowns can be used to control a subject without resorting to deadly force. Marines must train to become proficient in nonlethal techniques and to respond in a responsible manner. These techniques are referred to as compliance techniques, and they are applied in the third level in the continuum of force.

WARNING

During training, never apply the techniques for unarmed restraints and manipulation at full force or full speed. Use a slow and steady pressure to avoid injury.

Compliance Techniques

Compliance techniques are unarmed restraint and manipulation techniques used to physically force a subject or opponent to comply. Compliance can be achieved through the close combat techniques of-

Pain compliance using joint manipulation and pressure points. (Pain compliance is the initiation of pain to get compliance on the part of the subject.)
Come-along holds.

Principles of Joint Manipulation

Joint manipulation is used to initiate pain compliance and gain control of a subject. It involves the application of pressure on thejoints (elbow, wrist, shoulder, knee, ankle, and fingers). Pressure is applied in two ways:

In the direction in which the joint will not bend. For example, joints such as the knees

and elbows only bend in one direction and when pressure is applied in the opposite direction, pain compliance can be achieved. Beyond the point where the joint stops naturally in its range of movement (i.e., it no longer bends).

Since each joint has a breaking point, Marines should apply slow steady pressure only until pain compliance is reached. Continued pressure will break the joint and may escalate the violence of the situation.

Joint manipulation also uses the principle of off-balancing. A subject can be better controlled when he is knocked off balance.

Wristlocks

A wristlock is a joint manipulation that can be applied in a number of ways to achieve pain compliance. The wrist rotates in a number of directions and will bend in a single direction until its movement stops naturally. In a wristlock, pressure is exerted beyond that point by bending or twisting the joint. A wristlock is executed when an opponent tries to grab Marines or is successful in grabbing Marines or their equipment. A wristlock can also be performed by Marines if they wish to initiate control of an opponent.

Basic Wristlock. A basic wristlock is executed when Marines grab the opponent's left hand with the right hand. To execute the basic wristlock, Marines—

Use the right hand to grab the opponent's left hand by placing the thumb on the back of the opponent's hand so that the Marine's knuckles are facing to the left.

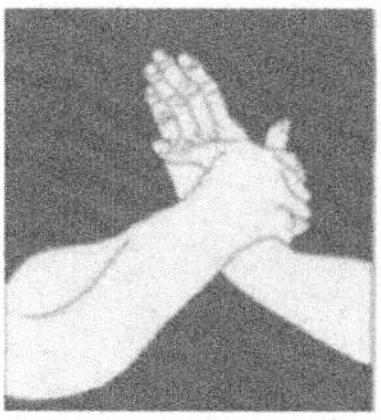

Hook the fingers across the fleshy part of the opponent's palm below the thumb. The fingers are used to anchor the hand so leverage can be applied to twist and bend the joint.

Exert downward pressure with the thumb to bend the opponent's joint. Rotate the opponent's hand to the right to twist the joint.

Step in to the opponent to keep the opponent's hand in close to the body to control him and provide more leverage on the wristlock.

Note: When executing the basic wristlock with the left hand, the Marine grabs the opponent's hand so that the Marine's knuckles are facing to the right, and then rotates and twists the opponent's hand to the left.

Use the left hand to further control the opponent.

Reverse Wristlock. A reverse wristlock is executed when Marines grab the opponent's right hand with the right hand. To execute the reverse wristlock, Marines—

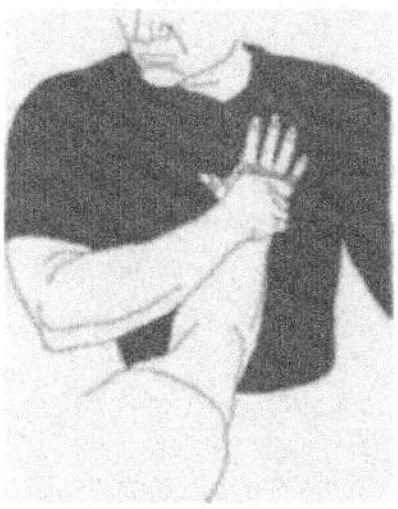

Place the right palm on the back of the opponent's right hand and wrap the fingers across the fleshy part of his palm below his little finger.

Twist the opponent's hand to the right while stepping in to place his hand against the chest. Apply downward pressure on the opponent's hand against the chest. Leave the opponent's hand on the chest to fully control the subject and to gain leverage.

Lean forward to use body weight to add additional pressure to the joint.

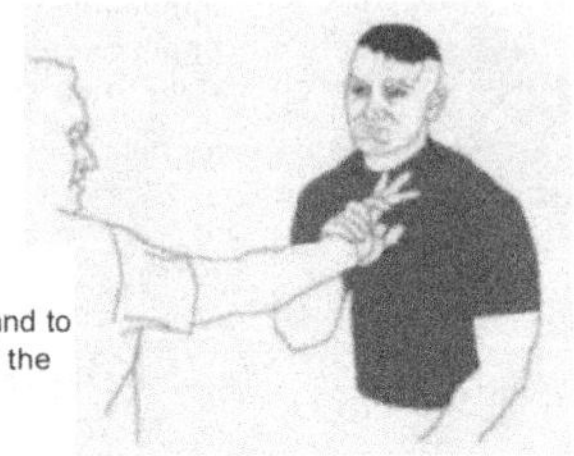

Use the left hand to further control the opponent.

Two-Handed Wristlock. Both hands can be used in the wristlock to maximize the leverage and pressure needed to bend and twist the joint. To execute the two-handed wristlock, Marines—

Place both thumbs on the back of the opponent's hand, thumbs crossed.

Hook the fingers of both hands around the fleshy part of the opponent's palm on both sides of his hand.

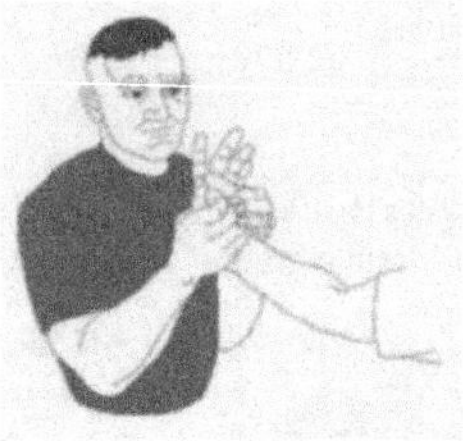

Step into the opponent and apply pressure downward on the back of his hand to bend the joint and rotate his wrist away from the body to twist the joint.

Enhanced Pain Compliance on Wristlock. Enhanced pain compliance techniques are applied in the third and fourth levels in the continuum of force. Additional pain be applied to a wristlock by-

Adding downward pressure to the elbow with the other hand or elbow by using the fingers to pull in on the opponent's radial nerve located on the inside of the forearm. When pressure is added to the opponent's radial nerve, his direction can be controlled.

Applying pressure against the opponent's finger joint to bend it in a direction it cannot bend (i.e., splitting the fingers).

Come-Along Holds

Marines use a come-along hold to control and move an opponent.

Escort Position. A common come-along hold is the escort position. To execute the escort position, Marines—

Face the opponent. Use the left foot to step forward at a 45-degree angle. Turn to face the right side of the opponent.

Use the right hand to firmly grasp the opponent's right wrist. With the left hand, firmly grasp the opponent's right triceps.

Position the opponent's controlled arm diagonally across the torso, keeping his wrist against the right hip. The Marine should be standing to the right of and behind the opponent.

Note: This technique works well when escorting an opponent on either the right or left side. Take caution when escorting an opponent by ensuring his controlled hand is not in a position to grab the holstered weapon. The preferred escort position is from the Marine's left side, so that the opponent is kept further away from the weapon.

Wristlock Come-Along. To execute the wristlock come-along, Marines-

Use the left hand to execute a basic wristlock. Incorporate the right hand in a two-handed wristlock for more control.

Maintain pressure on the opponent's wrist with the right hand, step forward, and pivot around to stand next to the opponent.

Release the left hand, quickly reach under the opponent's arm from behind, and grab his hand.

Use the left hand and apply downward pressure on the opponent's wrist.

Controlling Technique. The following controlling technique is used when an opponent grabs the Marine's wrist. To execute the technique, Marines—

Trap the opponent's hand with the palm of the other hand.

Rotate the opponent's trapped hand up and on his forearm while maintaining downward pressure on his trapped hand.

Apply downward pressure with both hands until the opponent is taken to the ground.

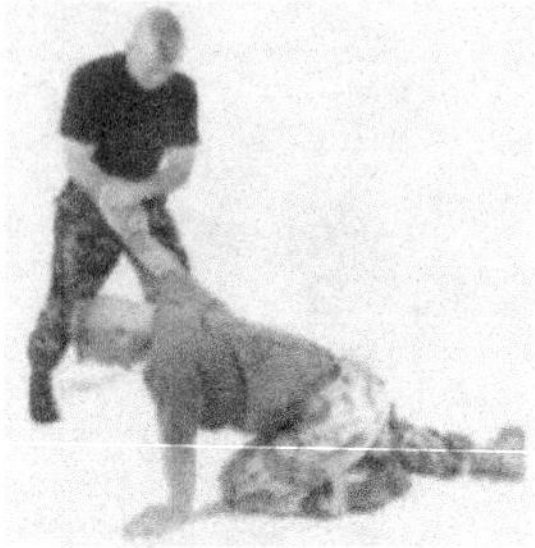

Armbars

An armbar is a joint manipulation in which pressure is applied on a locked elbow, just above the joint, in the direction the joint will not bend. An armbar has to be locked in quickly, but still requires a slow, steady pressure to gain compliance.

Basic Armbar. To execute a basic armbar, Marines-

Use the right hand to grab the opponent's right wrist.

Bring the left hand down on or above the opponent's elbow joint. To gain additional leverage, pivot to face the opponent.

Use the left hand to apply downward pressure on or above the opponent's elbow joint while pulling up on his wrist.

Armbar from a Wristlock. To execute an armbar from a wristlock, Marines-

Use the right hand to grab the opponent's right hand and execute a reverse wristlock.

Bring the left hand down on or above the opponent's elbow joint.

Use the left hand to apply downward pressure on or above the opponent's elbow joint while pulling up on his wrist.

Takedowns

A takedown is used to bring an opponent to the ground to further control him.

Takedown From a Wristlock Come-Along. To take the opponent to the ground from a wristlock come-along, Marines—

Use the right foot to push down on the opponent's calf or Achilles tendon.

Maintain control of the opponent's wrist and elbow and apply a slow, steady pressure to bring him to the ground.

Armbar to a Takedown. This technique is used to take a noncompliant opponent to the ground from an armbar. To execute the armbar takedown, Marines—

Use the right hand to execute a reverse wristlock.

Bring the left hand or forearm down on or above the opponent's elbow joint.

Pivot so the back is facing the opponent and, at the same time, lift the left elbow and slide the body so it is against the opponent, placing the armpit high above the opponent's elbow joint.

Lean back, placing the body weight on the opponent's arm until he complies or is taken to the ground.

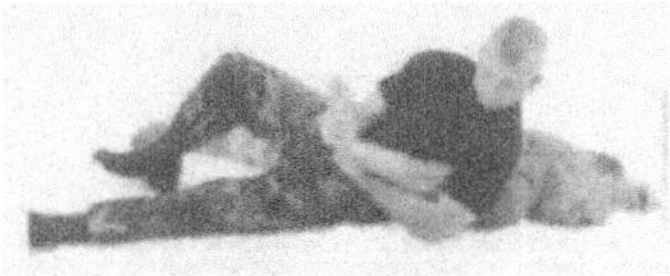

Note: This technique may break the opponent's arm. Therefore, this technique should not be employed if the objective is a nonlethal takedown.

Wristlock Takedown. This technique is used to take a noncompliant opponent to the ground from a basic wristlock and to put him in a position where he can be handcuffed, if necessary. To execute the takedown, Marines-

Use the right hand to execute a basic wristlock.

Incorporate the left hand in a two-handed wristlock.

Apply downward pressure on the wristlock, pivot on the ball of the right foot, and quickly turn to the right to take the opponent to the ground.

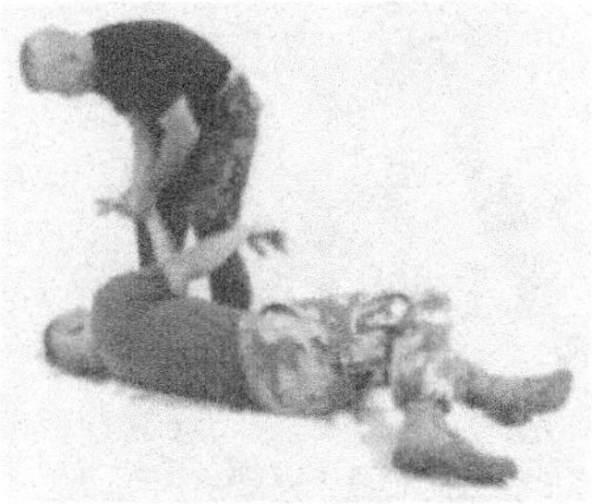

Continue to apply pressure on the wristjoint as the opponent lands on his back with his arm straight in the air.

Slide the left foot under the opponent's back, underneath his armpit.

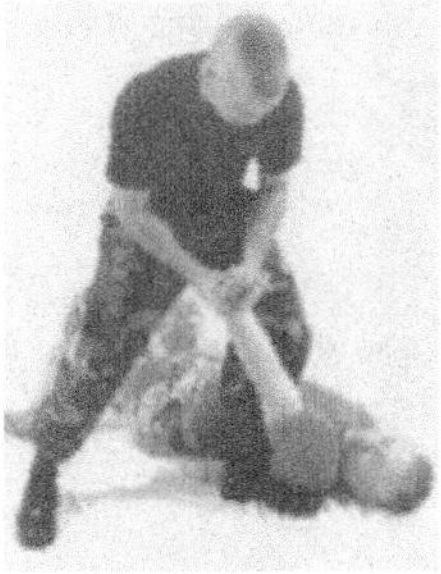

Apply pressure with the knee against the opponent's triceps while pulling back on his arm.

Continue applying pressure with the knee on the opponent's arm. Pivot and step around the opponent's arm to roll him on his stomach.

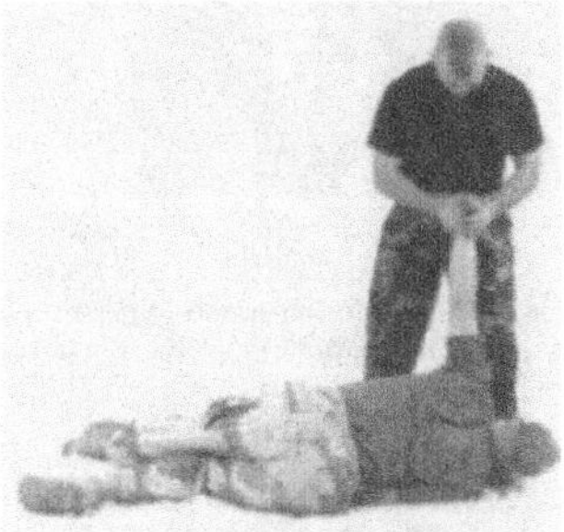

Kneel down with one knee on the opponent's back. The other knee is placed on the opponent's neck and shoulder on either side of his arm. Apply inward pressure with the knees to lock his arm in place.

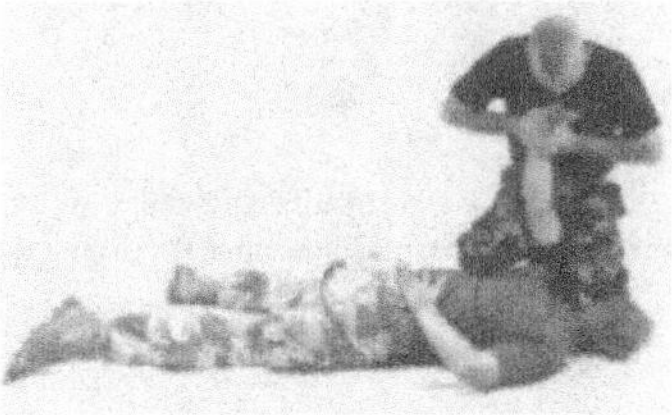

Tell the opponent to put his other hand in the middle of his back. Bring the opponent's controlled hand to the center of his back.

Escort Position Takedown. This technique can be used to control a noncompliant opponent from the escort position. To execute the escort position takedown, Marines—

Lock the opponent's arm straight across the body while rotating his wrist away from the body.

Use the left hand or forearm and apply downward pressure above the opponent's elbow where the triceps meet.

Step back with the right foot and, keeping the opponent's hand controlled against the hip, pivot to the right while continuing to apply downward pressure on his arm to bring him to the ground.

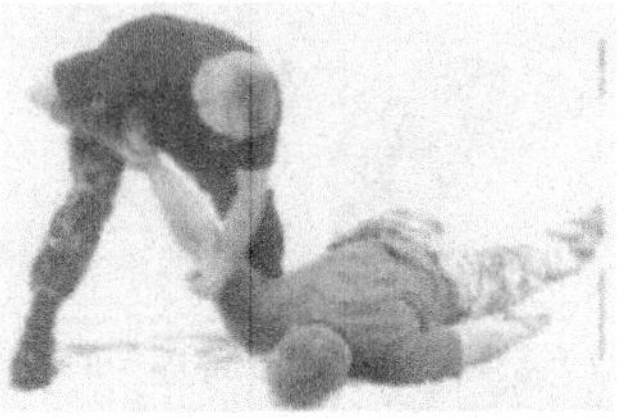

Note: This technique works well when escorting a subject on the Marine's right or left side. When taking down a subject from the right side, step back and pivot to the left.

2. Nonlethal Baton

A baton or nightstick can be an effective compliance tool when used correctly. Batons can be used defensively (blocking), offensively (striking), and as a restraining device when needed. In the fourth level of the continuum of force (assaultive [bodily harm]), defensive tactics include baton or nightstick blocks and blows. Blows to the head or other bony parts of the body are considered deadly force. When deadly force is not authorized, Marines must be able to employ blocks, strikes, and restraints with the baton with the minimum of force.

Grips

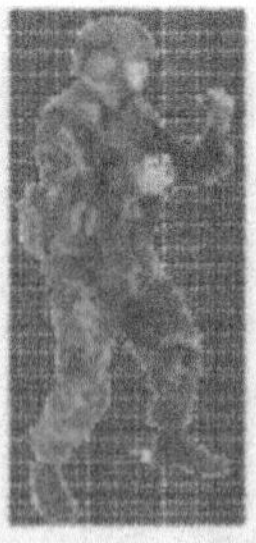

One-Handed Grip. To execute the one-handed grip, Marines—

Use the right hand to grasp the lower end of the baton, about 2 inches from the end.

Wrap the thumb and index finger around the baton so they are touching one another. The grip on the baton should be firm, but natural.

Two-Handed Grip. To execute the two-handed grip, Marines-

Use the right hand to grasp the lower end of the baton, about 2 inches from the end. Wrap the thumb and index finger around the baton so they touch one another.

Use the left hand to grasp the upper end of the baton, palm down, about 2 inches from the end. The hands should be approximately 10 to 12 inches apart.

Stance and Method of Carry

The basic warrior stance serves as the foundation for initiating nonlethal baton techniques. The method of cany provides effective defensive positions with a wide range of options to control a combative opponent.

One-Handed Carry. To execute the one-handed carry, Marines—

Grip the baton using the one-handed grip.

Elevate the baton, with the gripping hand at a level between the belt and shoulder.

Keep the left hand in the position of the basic warrior stance.

Two-Handed Carry. This carry is effective for blocks. To execute the two-handed carry, Marines-

Grip the baton using the two-handed grip.

Elevate the baton, with the left hand higher than the right hand.

Orient the weapon toward the subject.

Movement

In a nonlethal confrontation, movement may be made to create distance between Marines and the opponent or to close the gap to control him. When facing an opponent, Marines move in a 45-degree angle to either side of the opponent. Moving at a 45-degree angle is the best way to both avoid an opponent's strike and to put Marines in the best position to control the opponent.

Blocking Techniques

One-handed blocks are used when carrying the baton in the one-handed carry. The same one-handed blocks used in combative stick techniques (see page 3-6) apply in nonlethal baton engagements. The blocks include blocks for a vertical strike, a forward strike, and a reverse strike. Because the baton is often carried with two hands, there are also two-handed blocks that are effectively used from this carry. Two-handed blocks are discussed in the following subparagraphs.

High Block. Marines execute a high block to deter a downward vertical attack directed at the head and shoulders. To execute the high block, Marines—

Raise the baton up to a level even with or above the head. The baton should be in a horizontal position to block the blow. The fingers of the left hand should be open and behind the baton.

Place the baton perpendicular to the opponent's striking surface to absorb the impact of the blow.

Bend the elbows slightly to help absorb the impact of the blow. The arms should give with the strike of the blow.

Low Block. Marines execute a low block to deter an upward vertical attack directed at the abdomen, groin, or torso. The opponent's blow can be delivered by a foot, knee, or fist. To execute the low block, Marines—

Lower the baton to a level even with or below the groin. The baton should be in a horizontal position to block the blow. The fingers of the left hand should be open and behind the baton.

Place the baton perpendicular to the opponent's striking surface to absorb the impact of the blow.

Bend the elbows slightly to help absorb the impact of the blow. The arms should give with the strike of the blow.

Right Block. Marines execute a right block to deter an attack directed at the head, neck, flank, or hip. The opponent's blow can be delivered by a foot, knee, fist, or elbow. To execute the right block, Marines-

Thrust the baton in a vertical position to the right side.

Pivot to the right by stepping with the left foot and pivoting off the ball of the right foot. Rotate the hips and shoulders into the direction of the block. The fingers of the left hand should be open and behind the baton.

Place the baton perpendicular to the opponent's striking surface to absorb the impact of the blow.

Bend the elbows slightly to help absorb the impact of the blow. The arms should give with the strike of the blow.

Left Block. Marines execute a left block to deter an attack directed at the head, neck, flank, or hip. The opponent's blow can be delivered by a foot, knee, fist, or elbow. To execute the left block, Marines—

Thrust the baton in a vertical position to the left side.

Pivot to the left by stepping with the right foot and pivoting off the ball of the left foot. Rotate the hips and shoulders into the direction of the block. The fingers of the left hand should be open and behind the baton.

Place the baton perpendicular to the opponent's striking surface to absorb the impact of the blow.

Bend the elbows slightly to help absorb the impact of the blow. The arms should give with the strike of the blow.

Middle Block. Marines execute either a left or right block to deter an attack directed at the face, throat, chest, or abdomen. To execute the middle block, Marines—

Thrust the baton in a vertical position straight out in front of the body. The fingers of the left hand should be open and behind the baton.

Place the baton perpendicular to the opponent's striking surface to absorb the impact of the blow. The baton should be held with the left hand forward of the right.

Bend the elbows slightly to help absorb the impact of the blow. The arms should give with the strike of the blow.

Restraining Technique

The strong-side armlock is used to restrain an opponent who is not compliant. To execute the strong-side armlock, Marines—

Use the right hand to run the baton up under the opponent's left armpit, parallel to the ground.

Use the right foot to step forward at a 45-degree angle to the left side of the opponent. The baton is across his forearm.

Use the right hand to drive the baton forward and up so the action bends the opponent's arm behind his back. At the same time, continue moving around him to get behind him.

Place the baton on the opponent's forearm with the thumb and/or fingers and apply pressure to his forearm. At the same time, grasp the other end of the baton with the left hand.

Pull up on the low end of the baton with the right hand. At the same time, push down on the top end of the baton with the left forearm, reaching around with the left hand to grasp the opponent's biceps or shoulder.

Continue exerting downward pressure with the left forearm while pulling back on the opponent's biceps with the left hand. This places the opponent in a position where he is controlled and can be moved.

Apply pressure with the foot against the bend in the opponent's leg above his calf. This lowers the opponent to the ground rather than throwing him to the ground and risking severe injury.

Striking Target Areas

Marines must avoid striking an opponent in the head, neck, or other bony parts with the baton because this is considered deadly force and can lead to serious bodily injury or death. Instead, the legs, arms, and buttocks are target areas that are considered nonlethal.

Legs. Primary targets are the thighs and lower legs. Avoid striking the knee and ankle joints because this can cause permanent damage.

Arms. Primary targets are the upper arms. Avoid striking the shoulder and elbow and wrist joints because this can cause permanent damage.

Buttocks. Primary targets are the buttocks. Avoid striking any other part of the torso, including the chest, rib cage, spine, tail bone, and groin because strikes to these areas can cause permanent damage or death.

One-Handed Striking Techniques

One-Handed Forward Strike. A forward strike follows either a horizontal line or a downward diagonal line using a forehand stroke. To execute the one-handed forward strike, Marines-
Stand facing the opponent with the baton carried in a one-handed cany.

Place the right hand palm up, swing the baton from right to left, and make contact with the opponent.

One-Handed Reverse Strike. A reverse strike follows either a horizontal line or a downward diagonal line using a backhand stroke. To execute the one-handed reverse strike, Marines—

Stand facing the opponent with the baton carried in a one-handed carry.

Bend the right arm and cross the arm to the left side of the body. The baton should be close to or over the left shoulder.

Place the right hand palm down, swing the baton from left to right, and make contact with the opponent.

Two-Handed Striking Techniques

Two-Handed Forward Strike. This strike is an effective follow-up to a middle block or left block. To execute the two-handed forward strike, Marines—

Pull back with the left hand while driving the right hand forward toward the opponent. The baton should be horizontal to the ground. Power is generated by stepping forward with the right foot and rotating the right hip and shoulder into the strike.

Contact the opponent with the end of the baton.

Two-Handed Reverse Strike. This strike is an effective follow-up to a middle block or left block. To execute the two-handed reverse strike, Marines—

Pull back with the right hand while driving the left hand forward toward the opponent. The baton should be horizontal to the ground. Power is generated by stepping forward slightly with the left foot and rotating the left hip and shoulder into the strike.

Contact the subject with the end of the baton.

Front Jab. This strike is effective for countering a frontal attack. It can also be executed as a quick poke to keep a subject away. To execute the front jab, Marines thrust both hands forward in a quick jab. The baton is held either horizontal to the ground or at a slight downward angle.

Rear Jab. This strike is effective for countering a bear hug from the rear. To execute the rear jab, Marines thrust both hands rearward in a quick jab. The baton is held either horizontal to the ground or at a slight downward angle.

APPENDIX A

PUGIL STICK TRAINING

A pugil stick is a training device used to simulate a rifle bayonet so that effective, but safe, training can be conducted to build proficiency of rifle bayonet techniques. Pugil stick training builds on the techniques used to throw punches. Pugil stick training is the only "full contact" training provided to Marines in the Close Combat Program. Pugil stick training teaches Marines to function when faced with stress and violence, and it prepares them to deliver a blow and take a blow. It also provides them with the physical and mental skills vital to success on the battlefield.

1. Pugil Stick Training

Design

A pugil stick consists of a stick wrapped in padding at both ends that can be gripped like a rifle. The pugil stick is approximately the same weight and length of an unloaded rifle with a bayonet attached.

Holding the Pugil Stick

The pugil stick is held in the same manner as the sen ice rifle. All movements come from the basic warrior stance. To hold the pugil stick correctly, Marines—

Use the right hand to grasp the lower end of the pugil stick overhanded.
Use the left hand to grasp the upper end of the pugil stick underhanded.
Use the right forearm to lock the lower end of the pugil stick against the hip.
Orient the blade end of the pugil stick toward the opponent.

Safety Equipment

The following safety equipment must be worn during any pugil stick bout.

Groin Protection. Groin protection protects the groin from an accidental blow. It should be pulled high around the waist to protect the groin area, with the concave portion against the body.

Flak Jacket. The flak jacket provides protection to the body. It is worn completely fastened.

Neck Roll. The neck roll prevents whiplash if Marines receive a blow to the head. The neck roll further supports the head and protects the neck from blows. The neck roll is worn above the flak jacket and below the helmet. The tied end of the neck roll faces the front.

Helmet. A regulation football helmet protects the face and head. The helmet must fit snugly and the chin straps must be adjusted and snapped.

Mouthpiece. The mouthpiece is worn on the upper teeth to protect the teeth.

Gloves. Marines wear gloves to protect the hands if pugil sticks do not have gloves built-in.

Pugil Stick Screening

Prior to pugil stick training, instructors must ask participating Marines the following questions. If a Marine answers "yes" to any question, he must see the corpsman or a competent medical authority who evaluates him and determines whether or not the he can participate in the training.

- Are you on light duty?
- Are you restricted to running shoes by a corpsman or a doctor?
- Have you fought in a pugil stick bout within the last 7 days?
- Have you received a blow to the head within the last 7 days?
- Have you had a concussion within the last 6 months?
- Have you had dental surgery within the last 24 hours?
- Do you have stitches or staples on your body?
- Have you had a shoulder or head injury within the last 5 years?
- Are you taking a prescription drug?
- Do you have an ear infection or current sinus infection?
- Have you had a broken bone within the last 6 months?

Safety Personnel

The following safety personnel are required to conduct pugil stick training:

- One close combat instructor must officiate the bout.
- One close combat instructor trainer, commissioned officer, or staff noncommissioned officer must be in the training area to serve as range safety officer.
- One corpsman must be in the training area.

Training Area

To prevent injury, Marines train on areas with soft footing (i.e., sand or grass). Training mats are not recommended because feet can stick to the mats, prohibiting movement or causing joint injuries by twisting a knee. Bouts should not take place on a hard surface area; e.g., a flight deck or parking lot. A boxing ring may be used to conduct pugil stick bouts; ring dimensions can vary as long as there is ample room to execute the techniques in the training area.

Second Impact Syndrome

Second Impact Syndrome occurs when a second blow to the head produces a second concussion that occurs within 1 week following a previous concussion (before recovery from the first concussion). Second Impact Syndrome causes rapid brain swelling and can cause death. Therefore, there must be 7 days between pugil stick bouts to reduce the risk of severe injury resulting from Second Impact Syndrome. The 7 day separation between pugil stick bouts significantly reduces the possibility of injury, particularly in someone who may have suffered a brain injury or concussion but shows no symptoms.

Any Marine who experiences headaches or the following symptoms after training must be examined by appropriate medical personnel:

- Blurred vision.
- Ringing in the ears.
- Dilation of the pupils.
- Slurred speech.
- Bleeding from ears or mouth.
- Swelling in head or neck area.
- Any unnatural discoloration in head or neck area.

The Marine should not be allowed to participate in pugil stick training or any other activity where a heavy blow might be sustained for a minimum of 7 days after the headache or other symptoms have subsided.

Safety Measures

The following safety measures must be followed:

- Gear must be worn properly throughout training.
- Contact lenses or glasses will not be worn.
- False teeth will be removed from the mouth.
- Nothing will be worn around the neck except the neck roll.

Competition among groups of Marines is authorized as long as it does not overshadow training objectives or compromise safety procedures.

Safety and proper techniques are paramount.

Safety is more important than competition.

2. General Rules and Regulations Governing Pugil Stick Bouts

Instructors and Support Personnel Requirements

At a minimum, one close combat instructor observes a pugil stick bout. For safety purposes, it is better to have two instructors judging a bout because each instructor can fully observe each of the fighters. The best position for observation is to the right of a fighter. This allows the instructor to see the fighter's facial expression and body movement. The instructor's position must not interfere with the fight.

In addition, one commissioned or staff noncommissioned officer (to serve as range safety officer) and one corpsman will be in the training area.

Prior to the Bout

Prior to the bout, Marines—

Are paired according to height, weight, and gender.

Wear the proper safety equipment.

Once Marines have properly donned the safety equipment, they wait for instructions from the close combat instructor. On the command of the close combat instructor, a pair of Marines will enter the ring. Once in the ring, the close combat instructor inspects each Marine for the proper safety equipment.

Note: There should he enough gear so that when two Marines are training, two other Marines can be donning safety equipment for the next bout.

During the Bout

The bout begins when the close combat instructor blows the whistle. All strikes are directed above the waist. Upon hearing a whistle blast, all fighting immediately ceases.

Stopping a Bout

There arc two reasons for stopping a bout: delivery of a scoring blow or an unsafe condition. The close combat instructor trainer, close combat instructor, or the range safety officer may stop the bout at any time an unsafe condition is observed.

Scoring Blow. A scoring blow is an offensive technique delivered to a vulnerable area of an opponent with sufficient force and precision to be considered as a disabling or killing blow. Scoring blows are not judged solely on the degree of force with which a blow is delivered, but on the accuracy and techniques employed. A scoring blow is defined as—

A straight thrust with the blade end of the weapon (red end of the pugil stick) to the opponent's face mask or throat.

A slash to the side of the opponent's helmet (below the ear) or neck with the red end of the pugil stick.

A heavy blow to the opponent's head with an authorized technique (i.e., buttstrokc, smash) using the butt (black end) of the pugil stick.

When a scoring blow is delivered, the close combat instructor blows the whistle to stop the bout.

Unsafe Condition. The bout will be stopped as soon as an unsafe condition exists. An unsafe condition exists when a Marine is unable to defend himself, loses his balance and falls down on one or two knees or falls down completely, shows instability (e.g., buckling at the knees), loses muscular tension in his neck and his head snaps back or to one side, or appears disoriented. An unsafe condition also exists if a Marine lets go of one end of the pugil stick, the equipment (e.g., helmet, neck roll) falls off, or a Marine fails to use the proper techniques.

If any of these conditions occur—

- The close combat instructor trainer, close combat instructor, range safety officer, or anyone supervising training stops the bout and separates the two Marines.
- The corpsman evaluates the possibly injured Marine to determine if the fight can continue. The corpsman—
  - Ensures that he is alert and responsive, both verbally and physically.
  - Talks to him to see if he is coherent.
  - Makes certain he comprehends and replies to verbal questions.
  - Makes certain his speech is not slurred.
  - Checks his physical signs.
  - Ensures his eyes are focused and not dazed or glazed.
  - Ensures his legs are not wobbly or shaky.

The pugil stick is not used as a baseball bat. The use of unauthorized techniques will result in expulsion from the bout. Only techniques taught in bayonet training are authorized. These techniques include-

- Straight thrust.
- Buttstroke (horizontal and vertical).
- Smash.
- Slash.
- Parry.
- Blocks (low, high, left, right).

After the Bout

A third whistle blast is used prior to identifying the Marine who delivered the killing blow.

A fourth whistle blast is used prior to demonstrating the killing technique used by the Marine who won the bout.

3. Directions for Making Pugil Sticks

Close combat instructor trainers are taught to make pugil sticks to standard specifications at school. It is recommended that all pugil sticks be made under the supervision of a close combat instructor trainer. Until the Marine Corps acquires a standardized pugil stick that units can order through the supply system, instructors use the following information and instructions to make pugil sticks locally.

Supplies

The following supplies are needed to make one pugil stick:

- One 45" by 2" circular, oak, wood dowel.
- Four 9" wide by 27" long foam pad pieces to make two inner handguards of double thickness.
- Four 11" wide by 19" long foam pad pieces to make two outer handguards of double thickness.
- Canvas material to cover inner and outer handguards.
- One foam pad cut 1/4" thick by 7" wide by 13" long to make the centerguard.
- One 8" wide by 14"long piece of canvas material to make a pugil stick sock.
- Canvas material to make two pugil stick end caps. Ideally, two different color fabrics should be used to simulate the bayonet and buttstock ends of a rifle. Common colors for end caps are black and red. Each end cap is a different color to help the officiating certified close combat instructor determine the scoring blow during the pugil stick bout.
- Two 1/4" thick by 11" wide by 54" long foam pad pieces to make two ends.
- Two foam pads cut 1/4" thick by 3" wide by 14 1/2" long to make two end plug inserts.
- One can of 3M Photo MountQ spray adhesive (NSN 8040-01-120-4009)to adhere foam padding to the stick. (Spray adhesive is available from self-service.)
- One roll of duct tape or riggers tape (NSN 7510-00-266-5016) to protect foam pad and

canvas from wear and to reinforce gluing action of spray adhesive. (Duct tape or riggers tape is available from self-service.)

One roll para-cord or 5-50 cord to tie both the inner and outer handguards together.

One roll of thread strong enough to sew canvas material together.

Sixteen grommets, one for each of the four corners of the inner and outer handguards.

One foam sleeping mat (NSN 8465-01-109-3369). Use unserviceable foam sleeping mats obtained from Defense Rcutilization and Marketing Office.

Tools

The follow ing tools are needed to make a pugil stick:

Tape measure, ruler, or yardstick.
Scissors.
Grommet machine.
Knife.
Sewing machine.

Directions

To make a pugil stick, Marines perform the following steps.

Preparing Materials

Step 1. Cut foam sleeping mat into the dimensions listed above. Cut mats to double thickness.

Step 2. Use spray adhesive to adhere the two 9" by 27" pieces of foam together. Use the spray adhesive to adhere the two 11" by 19" pieces of foam together. Let adhesive dry before using the double-thick foam for handguards.

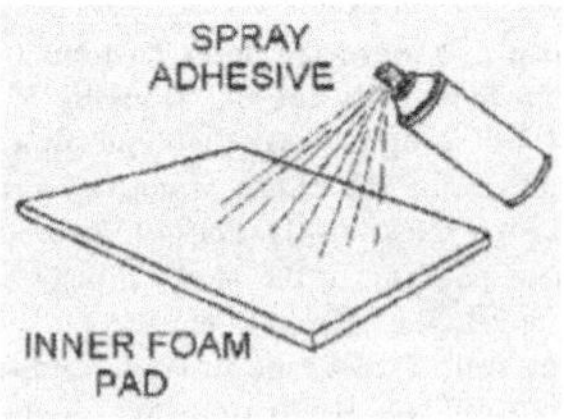

Step 3. Cut canvas material to make two coverings for the inner handguards. Cut material to make two coverings for the outer handguards. Two inner handguards and two outer handguards are required for each pugil stick.

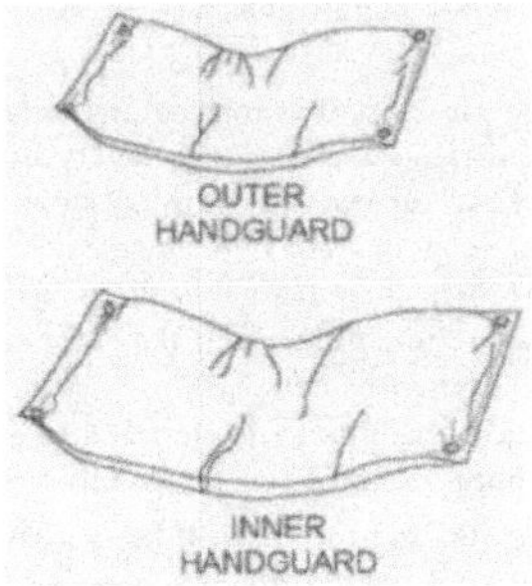

Step 4. Use a sewing machine to make a sleeve out of the 9" by 27" canvas. Sew down three sides. Insert foam pad inside the canvas sleeve and then sew down the fourth side. Use the grommet machine to create holes reinforced with metal grommets in each of the four corners. Once finished, the inner handguards arc complete.

Note: It is recommended that the Fabric Repair Shop be used to sew canvas sleeves and apply grommets.

Step 5. Repeat step 4 to make the other inner handguard and both outer handguards.

Constructing Ends, End Plugs, and End Caps

Step 1. Apply spray adhesive on approximately 8 inches of one end of the dowel. This adhesive serves to glue the foam pad to the wooden dowel.

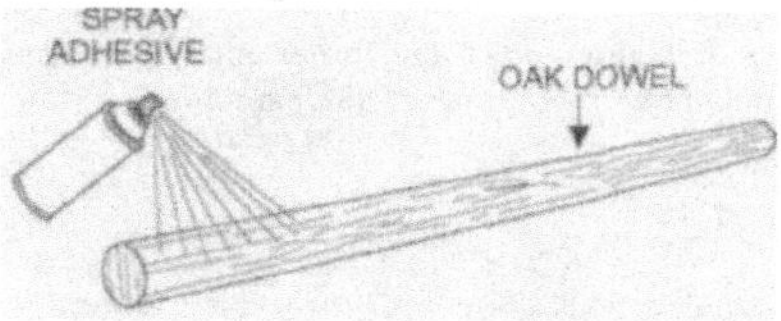

Step 2. Apply spray adhesive on one side of the 11" by 54" foam pad. This causes the foam pad to stick to itself when it is rolled around the end of the wooden dowel. Ensure 3 inches of the rolled foam pad extends past the wooden dowel for stability.

Step 3. Apply spray adhesive on one side of 3" by 14 1/2" foam pad. Roll pad into a tight cylinder.

Step 4. Apply spray adhesive on the exposed end of the dowel and fill the hole in the end with the end plug. The end plug prevents the wooden dowel from protruding beyond the edge of the rolled foam padding. This is an important safety precaution to prevent injury.

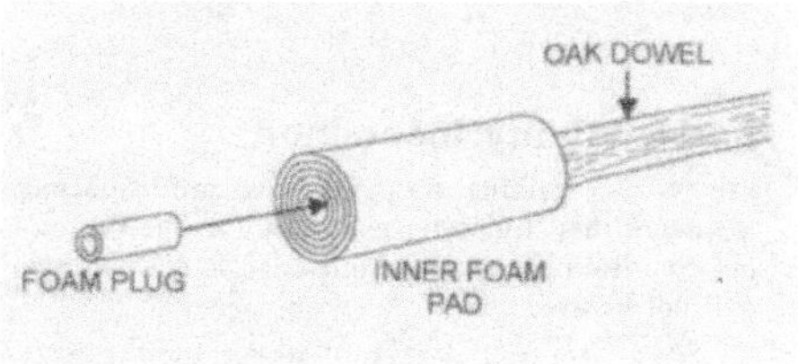

Step 5. Use duct tape or riggers tape to secure and reinforce the foam pad once the end plug is inserted in the end.

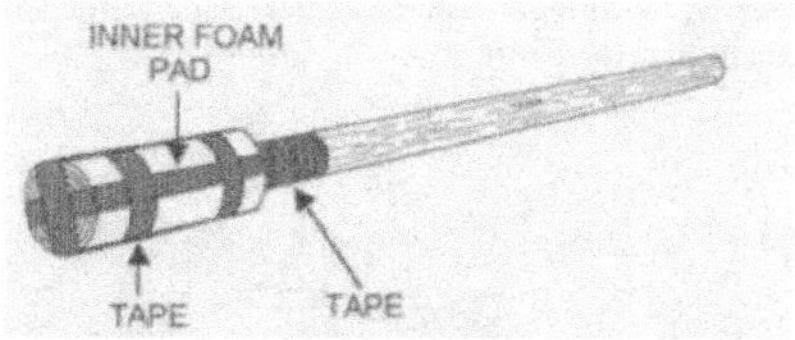

Step 6. Measure the dimensions of the ends (typically, the length of the cylinder should be 12 inches long to allow for taping). Cut and sew canvas material in the shape of a cylinder with one open end. Once complete, this becomes the end cap. The end cap is a colored sleeve that covers the rolled and taped foam pad at each end of the pugil stick.

Note: It is recommended that the Fabric Repair Shop be used to make the end cap.

Step 7. Slide one of the end caps over the rolled foam pad and tape it to the dowel.

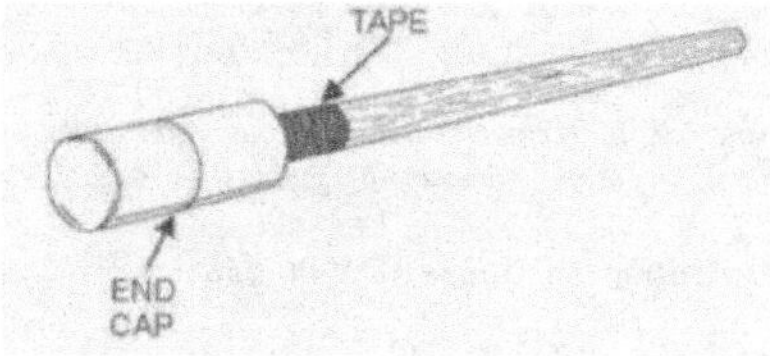

Building the Centerguard

Step 1. Spray one side of the 7" by 13" foam pad with spray adhesive and wrap it tightly around the center of the dowel.

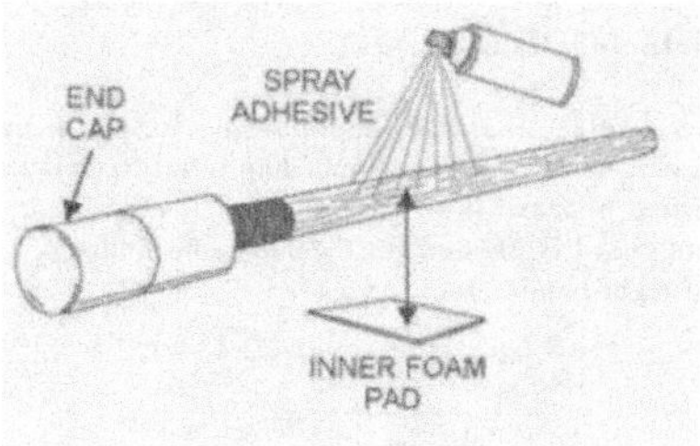

Step 2. Wrap duct tape or riggers tape around the foam pad to secure.

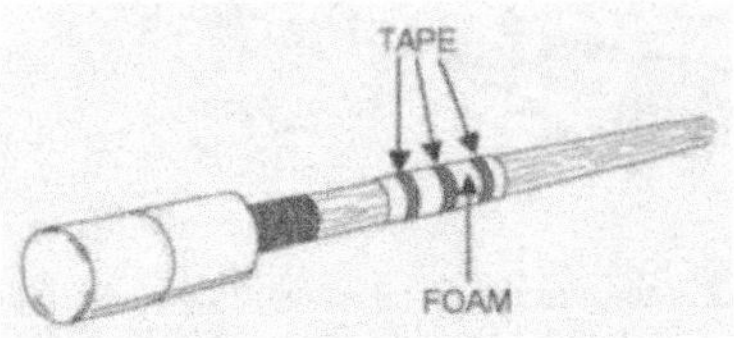

Step 3. Use a sewing machine to make a tubular pugil stick sock out of the 8" by 14" piece of canvas. Slide the pugil stick sock over the rolled foam pad and secure it with the duct tape or riggers tape. Duct tape or riggers tape protects the edges of the centerguard and reinforces the gluing effects of the spray adhesive.

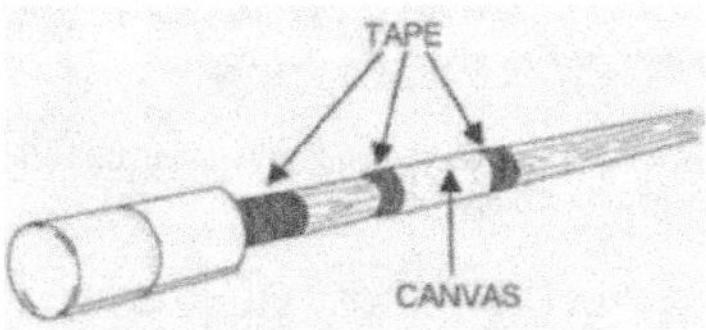

Note: It is recommended that the Fabric Repair Shop be used to make the pugil stick sock.

Finishing the Opposite End Cap

After one end cap is complete and the centerguard is finished, then repeat steps 1 through 7 in Constructing Ends, End Plugs, and End Caps for the other end of the pugil stick. Ensure that the opposite color end cap is used.

Fastening Handguards

Step 1. Place the inner handguards (longer piece) around the dowel. The inner handguard is placed on first because it protects the wrist. One inner handguard is needed for the left hand and one for the right hand.

Step 2. Fold the inner handguard around the pugil stick and run the para-cord or 5-50 cord through the metal grommets.

Step 3. Tighten down the corner ends of the inner handguards by pulling on the para-cord or 5-50 cord.

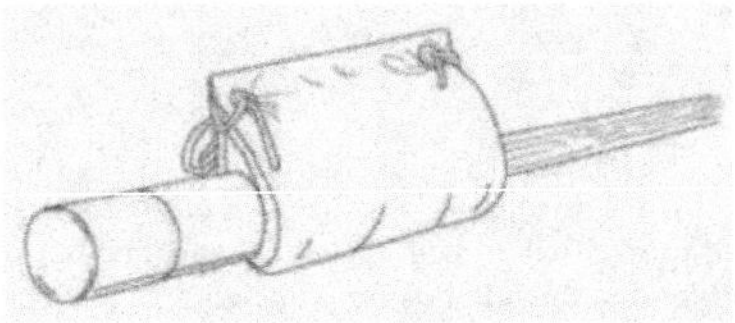

Step 4. Tie off the cord with a square knot above the metal grommets.

Step 5. Place the outer handguards (shorter piece) over the inner handguards (longer piece). Repeat steps 1-4 for the other handguard. The outer handguards provide additional protection to the knuckles and fingers and provides additional stability for the inner handguards.

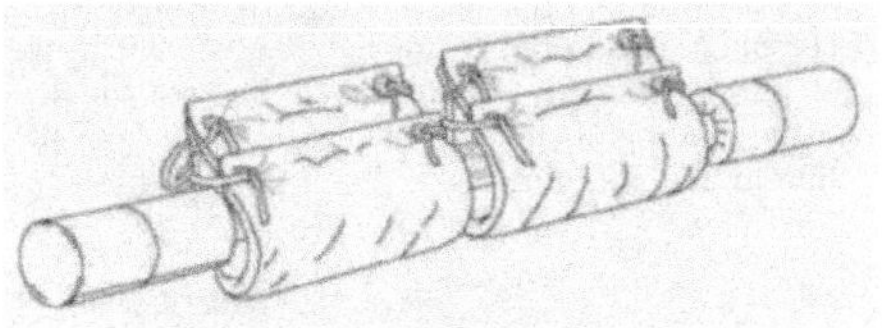

Serviceability Inspection

Prior to any training, all pugil sticks are inspected to ensure they are fit for use. If any of the following conditions exist, the pugil stick is unsafe and will not be used:

Stick protrudes through the end caps.
Frayed or worn edges exist on canvas surfaces.
The stick can be felt through the pads or the padding is too soft.
End caps or handguards are not securely attached.

APPENDIX B

SAFETY PRECAUTIONS DURING TRAINING

1. General Safety Precautions

When training close combat techniques, certain safety precautions must be adhered to to prevent injuries.

Most training should be conducted on a training area with soft footing such as a sandy or grassy area. If training mats are available, they should be used. A hard surface area is not appropriate for close combat training.

All techniques should be executed slowly at first. Marines can increase the speed of execution as they become more proficient. Marine-on-Marine training that requires contact (chokes, throws, ground fighting, and unarmed restraints and manipulation) should not be executed at full force or full speed.

If a technique is applied to the point that a Marine is uncomfortable, the Marine must "tap out." This indicates immediate release of the pressure being applied or to immediately stop the technique. The Marine "taps out" by firmly tapping his hand several times on any part of the opponent's body that will get his attention or by saying *stop*.

Second Impact Syndrome occurs when a second concussion develops within hours, days, or weeks following a prior concussion (and before recovery from the first concussion). Second Impact Syndrome causes rapid brain swelling and can cause death. Marines who experience headaches or other symptoms following training must be examined by appropriate medical personnel. These symptoms can include, but are not limited to, blurred vision, ringing in the ears, dialation of the pupils, bleeding from the ears or mouth, slurred speech, swelling in head or neck area, or any unnatural discoloration of head or neck. They should not be allowed to participate in pugil stick training or any other activity where a heavy blow might be sustained for a minimum of 7 days after the symptoms have subsided.

2. Safety Precautions for Individual Techniques

Falls

When training Marines to fall, they should progress from the ground, to a kneeling or squatting position, and then to a standing position. This ensures they are comfortable and understand the technique before progressing to executing falls from a higher profile. This instructional technique vastly reduces the risk of injury.

Strikes and Punches

When training Marines to strike and punch, they begin by executing the techniques "in the air." As they become more proficient, they execute strikes on equipment (when available) such as an air shield or a heavy bag. At no time should they be permitted to execute strikes on another student.

Chokes and Ground Fighting Chokes

When training Marines to execute chokes, they will not apply pressure to the opponent's throat during training because the trachea and windpipe can be crushed. During training, Marines should practice the proper procedures for blood chokes. They should never execute air chokes. At no time during training should a choke be applied at full force or full speed. No choke should be held for longer than 5 seconds.

Unarmed Restraints and Manipulation

When training Marines to execute unarmed restraints and manipulation techniques, they utilize slow and steady pressure. Never apply these techniques at full force or at full speed.

Pugil Stick Training

A blow to the head during training will bruise the brain. A second blow to the head can cause death. This is known as the Second Impact Syndrome. There must be 7 days between pugil stick training to prevent injury or death.

Mountain Operations

Department of Defense

Pentagon
Publishing

Preface

FM 3-97.6 describes the tactics, techniques, and procedures that the United States (US) Army uses to fight in mountainous regions. It is directly linked to doctrinal principles found in FM 3-0 and FM 3-100.40 and should be used in conjunction with them. It provides key information and considerations for commanders and staffs regarding how mountains affect personnel, equipment, and operations, it also assists them in planning, preparing, and executing operations, battles, and engagements in a mountainous environment.

Army units do not routinely train for operations in a mountainous environment. Therefore, commanders and trainers at all levels should use this manual in conjunction with TC 90-6-1. Army Training and Evaluation Program (ARTEP) mission training plans, and the training principles in FM 7-0 and FM 7-10 when preparing to conduct operations in mountainous terrain.

The proponent of this publication is Headquarters TRADOC. Send comments and recommendations on DA Form 2028 directly to Commander, US Army Combined Arms Center and Fort Leavenworth, ATTN: ATZL-SWW. Fort Leavenworth, Kansas 66027-6900.

Unless this publication states otherwise, masculine nouns and pronouns do not refer exclusively to men.

Field Manual
No. 3-97.6

***FM 3-97.6**

Headquarters
Department of the Army
Washington, DC, 28 November 2000

Mountain Operations

Contents

DISTRIBUTION RESTRICTION: Approved for public release; distribution unlimited

This publication supersedes FM 90-6, 30 June 1980.

Introduction

The US Army has a global area of responsibility and deploys to accomplish missions in both violent and nonviolent environments. The contemporary strategic environment and the scope of US commitment dictate that the US Army be prepared for a wide range of contingencies anywhere in the world, from the deserts of southwest Asia and the jungles of South America and southeast Asia to the Korean Peninsula and central and northern Europe. The multiplicity of possible missions makes the likelihood of US involvement in mountain operations extremely high. With approximately 38 percent of the world's landmass classified as mountains, the Army must be prepared to deter conflict, resist coercion, and defeat aggression in mountains as in other areas.

Throughout the course of history, armies have been significantly affected by the requirement to fight in mountains. During the 1982 Falkland Islands (Malvinas) War, the first British soldier to set foot on enemy-held territory on the island of South Georgia did so on a glacier. A 3.000-meter (10,000-foot) peak crowns the island, and great glaciers descend from the mountain spine. In southwest Asia, the borders of Iraq, Iran, and Turkey come together in mountainous terrain with elevations of up to 3,000 meters (10,000 feet).

Mountainous terrain influenced the outcome of many battles during the Iran-Iraq war of the 1980s. In the mountains of Kurdistan, small Kurdish formations took advantage of the terrain in an attempt to survive the Iraqi Army's attempt to eliminate them. In the wake of the successful United Nations (UN) coalition effort against Iraq, US forces provided humanitarian assistance to Kurdish people suffering from the effects of the harsh mountain climate.

Major mountain ranges, which are found in desert regions, jungles, and cold climate zones, present many challenges to military operations. Mountain operations may require special equipment, special training, and acclimatization. Historically, the focus of mountain operations has been to control the heights or passes. Changes in weaponry, equipment, and technology have not significantly shifted this focus. Commanders should understand a broad range of different requirements imposed by mountain terrain, including two key characteristics addressed in this manual: (1) the significant impact of severe environmental conditions on the capabilities of units and their equipment, and (2) the extreme difficulty of ground mobility in mountainous terrain.

Field Manual
No. 3-97.6

***FM 3-97.6**

Headquarters
Department of the Army
Washington, DC, 28 November 2000

Mountain Operations

Contents

DISTRIBUTION RESTRICTION: Approved for public release; distribution unlimited

This publication supersedes FM 90-6, 30 June 1980.

Chapter 1

Intelligence

Before they can understand how to fight in mountainous environment, commanders must analyze the area of operations (AO), understand its distinct characteristics, and understand how these characteristics affect personnel and equipment. This chapter provides detailed information on terrain and weather necessary to conduct a thorough intelligence preparation of the battlefield (IPB). however, the IPB process remains unaffected by mountains (see FM 2-01.3 for detailed information on how to conduct IPB).

SECTION I - THE PHYSICAL ENVIRONMENT

1-1. The requirement to conduct military operations in mountainous regions presents commanders with challenges distinct from those encountered in less rugged environments and demands increased perseverance, strength. will, and courage. Terrain characterized by steep slopes, great variations in local relief, natural obstacles, and lack of

CONTENTS

accessible routes restricts mobility, drastically increases movement times, limits the effectiveness of some weapons, and complicates supply operations. The weather, variable with the season and time of day, combined with the terrain, can greatly affect mobility and tactical operations. Even under non-violent conditions, operations in a mountainous environment may pose significant risks and dangers.

TERRAIN

1-2. Mountains may rise abruptly from the plains to form a giant barrier or ascend gradually as a series of parallel ridges extending unbroken for great distances. They may consist of varying combinations of isolated peaks, rounded crests, eroded ridges, high plains cut by valleys, gorges, and deep ravines. Some mountains, such as those found in desert regions, are dry and barren, with temperatures ranging from extreme heal in the summer to extreme cold in thewinter. In tropical regions, tush jungleswith heavy seasonal rains and little temperature variation frequently cover mountains. High, rocky crags with glaciated peaks and year-round snow cover exist in mountain ranges at most latitudes along the western portion of the Americas and in Asia. No matter what form mountains take, their common denominator is rugged terrain.

MOUNTAINOUS REGIONS

1-3. The principal mountain ranges of the world lie along the broad belts shown in Figure 1-1. Called Cordillera, after the Spanish word for rope, they encircle the Pacific basin and then lead westward across Eurasia into North Africa. Secondary, though less rugged, chains of mountains lie along the Atlantic margins of America and Europe.

Figure 1-1. Mountain Regions of the World

1-4. A broad mountainous region approximateiy 1,600 kilometers wide dominates northwestern North America. It occupies much of Alaska, more than a quarter of Canada and the US, and all but a small portion of Mexico and Central America The Rocky Mountain Range includes extensive high plains and basins. Numerous peaks in this belt rise above 3,000 meters (10,000 feet). Its climate varies from arctic cold to tropical heat, with the full range of seasonal and local extremes.

1-5, Farther south, the Andes stretch as a continuous narrow band along the western region of South America, Narrower than its counterpart in the north, this range is less than 800 kilometers wide. However, it continuously exceeds an elevation of 3,000 meters (10,000 feet) for a distance of 3,200 kilometers.

1-6. I n its western extreme, the Eurasian mountain belt includes the Pyrenees, Alps, Balkans, and Carpathian ranges of Europe. These loosely linked systems are separated by broad low basins and are cut by numerous valleys. The Atlas Mountains of North Africa are also a part of this belt Moving eastward into Asia, this system becomes more complex as it reaches the extreme heights of the Hindu Kush and the Himalayas. Just beyond the Pamir Knot on the Russian-Afghan frontier, it begins to fan out across all parts of eastern Asia. Branches of this belt continue south along the rugged island chains to New Zealand and northeast through the Bering Sea to Alaska.

MOUNTAIN CHARACTERISTICS

1-7. Mountain slopes generally vary between 15 and 45 degrees. Cliffs and other rocky precipices may be near vertical, or even overhanging. Aside from obvious rock formations and other local vegetation characteristics, actual slope surfaces are usually found as some type of relatively firm earth or grass. Grassy slopes may include grassy clumps known as tussocks, short alpine grasses, or tundra (the latter more common at higher elevations and latitudes). Many slopeswill be scattered with rocky debris deposited from the higher peaks and ridges. Extensive rock or boulder fields are known as talus. Slopes covered with smaller rocks, usually fist-si zed or smaller, are called scree fields. Slopes covered in talus often prove to be a relatively easy ascent route. On the other hand, climbing a scree slope can be extremely difficult, as the small rocks tend to loosen easi ly and giveway. H owever, this characteristic often makes scree fields excel lent descent routes. Before attempt i ng to descend scree slopes, commanders should carefully analyze the potential for creati ng dangerous rock fall and take necessary avoidance measures.

1-8. In winter, and at higher elevations throughout the year, snow may blanket slopes, creating an environment with its own distinct affects. Some snow conditions can aid travel by covering rough terrain with a consistent surface. Deep snow, however, greatly impedes movement and requires soldiers well-trained in using snowshoes, skis, and over-snow vehicles. Steep snow covered terrain presents the risk of snow avalanches as well. Snow can pose a serious threat to soldiers not properly trained and equipped for movement under such conditions. Avalanches have taken the lives of more soldiers engaged in mountain warfare than all other terrain hazards combined,

1-9. Commanders operating in arctic and subarctic mountain regions, as well as the upper elevations of the world's high mountains, may be confronted with vast areas of glaciation. Valleys in these areas are frequently buried under massive glaciers and present additional hazards, such as hidden crevices and ice and snow avalanches. The mountain slopes of these peaks are often glacialed and their surfaces are generally composed of varying combinations of rock, snow, and ice. Although glaciers have their own peculiar hazards requiring special training and equipment, movement over valley glaciers is often the safest route through these areas (TC 90-6-1 contains more information on avalanches and glaciers, and their effects on operations).

MOUNTAIN CLASSIFICATIONS

1-10. There is no simple system available to classify mountain environments. Soil composition, surface configuration, elevation, latitude, and climatic patterns determine the specific characteristics of each major mountain range. When alerted to the potential requirement to conduct mountain operations, commanders must carefully analyze each of these characteristics for the specific mountain region in which their forces will operate. However, mountains are generally classified or described according to their local relief; for military purposes, they may be classified according to operational terrain levels and dismounted mobility and skill requirements.

Local Relief

1-11. Mountains are commonly classified as low or high, depending on their local relief and, to some extent, elevation. Low mountains have a local relief of 300 to 900 meters (1,000 to 3,000 feet) with summits usually below the limberline. High mountains have a local relief usually exceeding 900 meters (3,000 feet) and are characterized by barren alpine zones above the timberline. Glaciers and perennial snow cover are common in high mountains and usually present commanders with more obstacles and hazards to movement than do low mountains.

Operational Terrain Levels

1-12. Mountain operations are generally carried out at three different operational terrain levels (see Figure 1-2). Level I terrain is located at the bottom of valleys and along the main lines of communications. At this level, heavy forces can operate, but maneuver space is often restricted. Light and heavy forces are normally combined, since vital lines of communication usually follow the valley highways, roads, and trails.

Level	Description
I	The bottoms of valleys and main lines of communications
II	The ridges, slopes, and passes that overlook valleys
III	The dominant terrain of the summit region

Figure 1-2. Operational Terrain Levels

1-13. Level II terrain lies between valleys and shoulders of mountains. Generally, narrow roads and trails, which serve as secondary lines of communication, cross this ridge system. Ground mobility is difficult and light forces will expend great effort on these ridges, since they can easily influence operations at Level I. Similarly, enemy positions at the next level can threaten operations on these ridges.

1-14. Level MI includes the dominant terrain of summit regions. Although summit regions may contain relatively gentleterrain, mobility in Level III is usually the most difficult to achieve and maintain Level III terrain, however, can provide opportunities for well-trained units to attack the enemy from the flanks and rear At this terrain level, acclimatized soldiers with advanced mountaineering training can infiltrate to attack lines of communication, logistics bases, air defense sites, and command infrastructures

Dismounted Mobility Classification

1-15. When conducting mountain operations, commanders must clearly understand the effect the operational terrain level has on dismounted movement. Therefore, in addition to the general mobility classification contained in FM 2-01.3 (unrestricted, restricted, severely restricted), mountainous terrain may be categorized into five classes based on the type of individual movement skill required (see Figure 1-3), Operations conducted in class 1 and 2 terrain require little to no mountaineering skills. Operations in class 3, 4, and 5 terrain require a higher level of mountaineering skills for safe and efficient movement. Commanders should plan and prepare for mountain operations based, in large part, on this type of terrain analysis.

Class	Terrain	Mobility Requirements	Skill Level Required*
1	Gentler slopes/ trails	Walking techniques	Unskilled (with some assistance) and Basic Mountaineers
2	Steeper/rugged terrain	Some use of hands	
3	Easy climbing	Fixed ropes where exposed	Basic Mountaineers (with assistance from assault climbers)
4	Steep/exposed climbing	Fixed ropes required	
5	Near vertical	Technical climbing required	Assault Climbers
* See Chapter 2 for a discussion of mountaineering skill levels			

Figure 1-3. Dismounted Mobility Classification

WEATHER

1-16. In general, mountain climates tend to be cooler, wetter versions of the climates of the surrounding lowlands. Most mountainous regions exhibit at least two different climatic zones — a zone at low elevations and another at elevations nearer the summit regions. In some areas, an almost endless variety of local dimates may exist within a given mountainous region. Conditions change markedly with elevation, latitude, and exposure to atmospheric winds and air masses. In addition, the climatic patterns of two ranges located at the same latitude may differ radically.

1-17. Like most other landforms, oceans influence mountain dimates. Mountain ranges in dose proximity to oceans and other large bodies of water usually exhibit a maritime dimate. Maritime dimates generally produce milder temperatures and much larger amounts of rain and snow. Their relatively mild winters produce heavy snowfalls, while their summer temperatures rarely get excessively hot. Mountains farther inland usually display a more continental climate Winters in this type climate are often bitterly cold, while summers can be extremely hot. Annual rain- and snowfall here is far less than in a maritime climate and may be quite scarce for long periods. Relatively shallow snow-packs are normal during a continental dimale's winter season.

1-18. Major mountain ranges force air masses and storm systems to drop significant amounts of rain and snow on the windward side of the range. Asair masses pass over mountains, the leeward slopes receive far less precipitation than the windward slopes. It is not uncommon for the dimate on the windward side of a mountain range to be humid and the dimate on the leeward side arid. This phenomenon affeds coastal mountains, as well as mountains farther inland. The deepest winter snow-packs will almost always be found on the windward side of mountain ranges. As a result, vegetation and forest characteristics may be markedly different between these two areas. Prevailing winds and storm patterns normally determine the severity of these effect s.

1-19. Mountain weather can be erratic, varying from strong winds to calm, and from extreme cold to relative warmth within a short time or a minor shift in locality. The severity and variance of the weather require soldiers to be prepared for alternating periods of heat and cold, as well as conditions ranging from dry to extremely wet. At higher elevations, noticeable temperature differences may exist between sunny and shady areas or between areas exposed to wind and those protected from it. This greatly increases every soldier's clothing load and a unit's overall logistical requirements. Figure 1-4 summarizes the effects of mountain weather discussed below. FM 2-33,201 and FM 3-97 22 contain additional information on how weather affects operation s,

TEMPERATURE

1-20. Normally, soldiers encounter a temperature drop of three to five degrees Fahrenheit per 300-meter (1,000-foot) gain in elevation. In an atmosphere containi ng considerable water vapor, the temperature drops about one degree Fahrenheit for every 100-meter (300-foot) increase. In very dry air. it drops about one degree Fahrenheit for every 50 meters (150 feet). However, on cold, dear, and calm mornings, when a troop movement or dimb begins from a valley, soldiers may encounter higher temperatures as they gain elevation This reversal of the normal situation is called temperature inversion Additionally, during winter months, the temperature is often higher during a storm than during periods of dear weather. However, the dampness of predpitation and penetration of the wind may still cause soldiers to chill faster This is compounded by the fad that the cover afforded by vegetation often does not exist abovethe trt?e-line. Under these conditions, com ma riders must weigh the tactical advantage of retaining positions on high ground against seeking shelter and warmth at lower elevations with reduced visibility.

1-21, At high elevations, there may be differences of 40 to 50 degrees Fahrenheit between the temperature in the sun and that in the shade. This is similar in magnitude to the day-to-night temperature fluctuations experienced in some deserts (see FM 3-97.3). Besides permitting rapid heating, the dear air at high altitudes also results in rapid cooling at night. Consequently. temperatures ri se swiftly after sunrise and drop quickly after sunset. M uch of the chilled air drains downward so that the differences between day and night temperatures ere greater in valleys than on slopes.

Weather Condition	Flat to Moderate Terrain Effects	Added Mountain Effects
Sunshine	Sunburn Snow blindness Temperature differences between sun and shade	Increased risk of sunburn and snow blindness Severe, unexpected temperature variations between sun and shade Avalanches
Wind	Windchill	Increased risk and severity of windchill Blowing debris or driven snow causing reduced visibility Avalanches
Rain	Reduced visibility Cooler temperatures	Landslides Flash floods Avalanches
Snow	Cold weather injuries Reduced mobility and visibility Snow blindness Blowing snow	Increased risk and severity of common effects Avalanches
Storms	Rain/snow Reduced visibility Lightning	Extended duration and intensity greatly affecting visibility and mobility Extremely high winds Avalanches
Fog	Reduced mobility/visibility	Increased frequency and duration
Cloudiness	Reduced visibility	Greatly decreased visibility at higher elevations

Figure 1-4. Comparison of Weather Effects

WIND

1-22. In high mountains, the ridges and passes are seldom calm. By contrast, strong winds in protected valleys are rare. Normally, wind velocity increases with altitude and is intensified by mountainous terrain. Valley breezes moving up-slope are more common in the morning, while descending mountain breezes are more common in the evening. Wind speed increases when winds are forced over ridges and peaks (orographic lifting), or when they funnel through narrowing mountain valleys, passes, and canyons (Venturi effect). Wind may blow with great force on an exposed mountainside or summit. As wind speed doubles, its force on an object nearly quadruples

1-23, Mountain winds cause rapid temperature changes and may result in blowing snow, sand, or debris that can impair movement and observation. Commanders should routinely consider the combined cooling effect of ambient temperature and wind (windchill) experienced by their soldiers (see Figure 1-5 on page 1-8). At higher elevations, air is considerably dryer than air at sea level. Due to this increased dryness, soldiers must increase their fluid intake by approximately one-third. However, equipment will not rust as quickly, and organic matter will decompose more slowly.

WIND SPEED		COOLING POWER OF WIND EXPRESSED AS "EQUIVALENT CHILL TEMPERATURE"																				
KNOTS	MPH	TEMPERATURE (° F)																				
CALM	CALM	40	35	30	25	20	15	10	5	0	-5	-10	-15	-20	-25	-30	-35	-40	-45	-50	-55	-60
		EQUIVALENT CHILL TEMPERATURE																				
3-6	5	35	30	25	20	15	10	5	0	-5	-10	-15	-20	-25	-30	-35	-40	-45	-50	-55	-60	-70
7-10	10	30	20	15	10	5	0	-10	-15	-20	-25	-35	-40	-45	-50	-60	-65	-70	-75	-80	-90	-95
11-15	15	25	15	10	0	-5	-10	-20	-25	-30	-40	-45	-50	-60	-65	-70	-80	-85	-90	-100	-105	-110
16-19	20	20	10	5	0	-10	-15	-20	-30	-35	-45	-50	-60	-65	-75	-80	-85	-90	-100	-110	-115	-120
20-23	25	15	10	0	-5	-15	-20	-30	-35	-45	-50	-60	-65	-75	-80	-90	-95	-105	-110	-120	-125	-135
24-28	30	10	5	0	-10	-20	-25	-30	-40	-50	-55	-65	-70	-80	-85	-95	-100	-110	-115	-125	-130	-140
29-32	35	10	5	-5	-10	-20	-30	-35	-40	-50	-60	-65	-75	-80	-90	-100	-105	-115	-120	-130	-135	-145
33-36	40	10	0	-5	-15	-20	-30	-35	-45	-55	-60	-70	-75	-85	-95	-100	-110	-115	-125	-130	-140	-150
WINDS ABOVE 40 HAVE LITTLE ADDITIONAL EFFECT		LITTLE DANGER					INCREASING DANGER (Flesh may freeze within 1 minute)						GREAT DANGER (Flesh may freeze within 30 secs)									

Figure 1-5. Windchill Chart

PRECIPITATION

1-24. The rapid rise of air masses over mountains creates distinct locaf weather patterns. Precipitation in mountains increases with elevation and occurs more often on the windward than on the leeward side of ranges. Maximum cloudiness and precipitation generally occur near 1,800 meters (6,000 feet) elevation in the middle latitudes and at lower levels in the higher latitudes. Usually, a heavily wooded belt marks the zone of maximum precipitation.

Rain and Snow

1-25. Both rain and snow are common in mountainous regions. Rain presents the same challenges as at lower elevations, but snow has a more significant influence on all operations. Depending on the specific region, snow may occur at anytime during the year at elevations above 1,500 meters (5,000 feet). Heavy snowfall greatly increases avalanche hazards and can force changes to previously selected movement routes. In certain regions, the intensity of snowfall may delay major operations for several months. Dry, flat riverbeds may initially seem to be excellent locations for assembly areas and support activities, however, heavy rains and rapidly thawing snow and ice may create flash floods many miles downstream from the actual location of the rain or snow.

Thunderstorms

1-26. Although thunderstorms are local and usually last only a short time, they can impede mountain operations. Interior ranges with continental climates are more conducive to thunderstorms than coastal ranges with maritime climates. In alpine zones, driving snow and sudden wind squalls often accompany thunderstorms. Ridges and peaks become focal points for lightning strikes, and the occurrence of lightning is greater in the summer than the winter. Although statistics do not show lightning to be a major mountaineering hazard, it should not be ignored and soldiers should take normal precautions, such as avoiding summits and ridges, water, and contact with metal objects.

Traveling Storms

1-27. Storms resulting from widespread atmospheric disturbances involve strong winds and heavy predpitation and are the most severe weather condition that occurs in the mountains. If soldiers encounter a traveling storm in alpine zones during winter, they should expect low temperatures, high winds, and blinding snow. These conditions may last several days longer than in the lowlands. Specific conditions vary depending on the path of the storm. However, when colder weather moves in, dearing at high elevations is usually slow.

1-28. The effects of fog in mountains are much the same as in other terrain. However, because of the topography, fog occurs more frequently in the mountains The high inddence of fog makes it a significant planning consideration as it restricts visibility and observation complicating reconnaissance and surveillance. However, fog may help fadlitate covert operations such as infiltration. Routes in areas with a high occurrence of fog may need to be marked and charted to fadlitate passage.

SECTION II - EFFECTS ON PERSONNEL

1-29. The mountain environment is complex and unforgiving of errors. Soldiers conducting operations anywhere, even under the best conditions, become cold, thirsty, tired, and energy-depleted. In the mountains however, they may become paralyzed by cold and thirst and incapacitated due to utter exhaustion. Conditions such as high elevations, rough terrain, and extremely unpredictable weather require leaders and soldiers who have a keen understanding of environmental threats and what to do about them.

1-30. A variety of individual soldier charaderistics and environmental conditions influence the type, prevalence, and severity of mountain illnesses and injuries (see Figure 1-6 on page 1-10). Due to combinations of these characteristics and conditions, soldiers often succumb to more than one illness or injury at a time, increasing the danger to life and limb. Three of the most common, cumulative, and subtle fadors affeding soldier ability under these variable conditions are nutrition (to indude water intake), decreased oxygen due to high altitude, and cold. Preventive measures, early recognition, and rapid treatment help minimize n embattle casualties due to these conditions (see Appendix A for detailed information on mountain-specific i 11n esses and injuries).

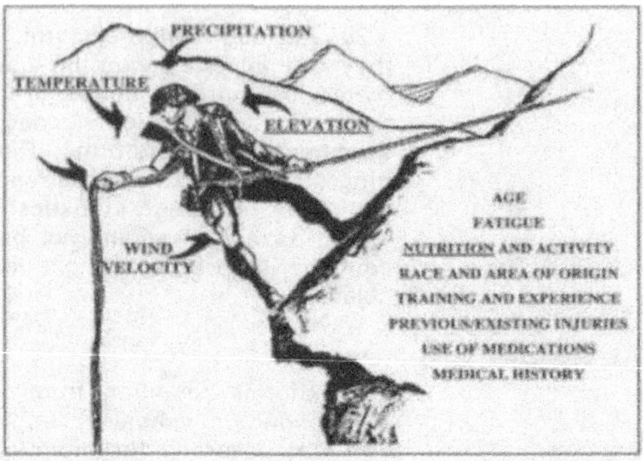

Figure 1-6. Environmental and Soldier Conditions Influencing Mountain Injuries and Illnesses

NUTRITION

1-31. Poor nutrition contributes to illness or injury, decreased performance, poor morale, and susceptibility to cold injuries, and can severely affect military operations. Influences at high altitudes that can affect nutrition include a dulled taste sensation (making food undesirable), nausea, and lack of energy or motivation to prepare or eat meals.

1-32, Caloric requirements increase in the mountains due to both the altitude and the cold A diet high in fat and carbohydrates is important in helping the body fight the effects of these conditions. Fats provide long-term, slew caloric release, but are often unpalatable to soldiers operating at higher altitudes. Snacking on high-carbohydrate foods is often the best way to maintain the calories necessary to function.

1-33. Products that can seriously impact soldier performance in mountain operations include:

- Tobacco. Tobacco smoke interferes with oxygen delivery by reducing the blood's oxygen-carrying capacity. Tobacco smoke in dose, confined spaces increases the amounts of carbon monoxide. The irritant effect of tobacco smoke may produce a narrowing of airways, interfering with optimal ai r movement. Smoking can effectively raise the"physiological altitude" as much as several hundred meters.
- Alcohol. Alcohol impairs judgement and perception, depresses respiration, causes dehydration, and increases susceptibility to cold injury
- Caffeine. Caffeine may improve physical and mental performance, but it also causes increased urination (leading to dehydration) and, therefore, should be consumed in moderation.

1-34. Significant body water is lost at higher elevations from rapid breathing, perspiration, and urination. Depending upon level of exertion, each soldier should consume about four to eight quarts of water or other decaffeinated fluids per day in low mountains and may need ten quarts or more per day in high mountains. Thirst is not a good indicator of the amount of water lost.

and in cold climates sweat, normally an indicator of loss of fluid, goes unnoticed. Sweat evaporates so rapidly or is absorbed so thoroughly by clothing layers that it is not readily apparent. When soldiersbecomelhirsty.they are already dehydrated. Loss of body water also plays a major role in causing altitude sickness and cold injury. Forced drinking in the absence of thirst, monitoring the deepness of the yellow hue in the urine, and watching for behavioral symptoms common to altitude sickness are important factors for commanders to consider In assessing the water balance of soldiers operating in the mountains.

1-35, In the mountains, as elsewhere, refilling each soldier's water containers as often as possible is mandatory. No matter how pure and clean mountain water may appear, water from natural sources should always be purified or chemically sterilized to prevent parasitical illnesses (giardiasis). Commanders should consider requiring the increased use of individual packages of powdered drink mixes, fruit, and juices to help encourage the required fluid intake.

ALTITUDE

1-36. As soldiers ascend in altitude, the proportion of oxygen in the air decreases. Without proper acclimatization, this decrease in oxygen saturation can cause altitude sickness and reduced physical and mental performance [see Figure 1-7). Soldiers cannot maintain the same physical performance at high altitudethat they can at low altitude, regardless of their fitness level.

Altitude	Meters	Feet	Effects
Low	Sea Level – 1,500	Sea Level – 5,000	None.
Moderate	1,500 – 2,400	5,000 – 8,000	Mild, temporary altitude sickness may occur
High	2,400 – 4,200	8,000 – 14,000	Altitude sickness and decreased performance is increasingly common
Very High	4,200 – 5,400	14,000 – 18,000	Altitude sickness and decreased performance is the rule
Extreme	5,400 – Higher	18,000 - Higher	With acclimatization, soldiers can function for short periods of time

Figure 1-7. Effects of Altitude

1-37. The mental effects most noticeable at high altitudes indude decreased perception, memory, judgement, and attention. Exposure to ait it udes of over 3,000 meters (10,000 feet) may also result in changes in senses, mood, and personality. Within hours of ascent, many soldiers may experience euphoria, joy, and excitement that are likely to be accompanied by errors i n judgement, leading to mistakes and accidents. After a period of about 6 to 12 hours, euphoria decreases, often changing to varying degrees of depression. Soldiers may become irritable or may appear listless. U sing the buddy system during this early exposure helps to identify soldiers who may be more severely affected. High morale and esprit instilled before deployment and reinforced frequently help to mini mize the i mpact of negative mood changes.

1-38. Trie physical effect most noticeable at high altitudes includes vision. Vision is generally the sense most affected by altitude exposure and can potentially affect military operations at higher elevations. Night vision is significantly reduced, affecting soldiers at approximately 2.400 meters (8,000 feet) or higher. Some effects occur early and are temporary, while others may persist after acclimatization or even for a period of time after descent. To compensate for loss of functional abilities, commanders should make use of tactics, techniques, and procedures that trade speed for increased accuracy. By allowing extra time lo accomplish tasks, commanders can minimize errors and injuries.

HYPOXIA-RELATED ILLNESSES AND EFFECTS

1-3g Hypoxia, a deficiency of oxygen reaching the tissues of the body, has been the cause of many mountain illnesses, injuries, and deaths. It affects everyone, but some soldiers are more vulnerable than others. A soldier may be affected at one time but not at another. Altitude hypoxia is a killer, but it seldom strikes alone. The combination of improper nutrition, hypoxia, and cold is much more dangerous than any of them alone. The three most significant altitude-related illnesses and their symptoms, which are essentially a series of illnesses associated with oxygen deprivation, are:

- Acute Mountain Sickness (AMS). Headache, nausea, vomiting, fatigue, irritability, and dizziness.

  High Altitude Pulmonary Edema (HAPE), Coughing, noisy breathing, wheezing, gurgling in the airway, difficulty breathing, and pink frothy sputum (saliva). Ultimately coma and death will occur without treatment,
- High Altitude Cerebral Edema (HACE). HACE is the most severe illness associated with high altitudes. Its symptoms often resemble AMS {severe headache, nausea, vomit i ng), often with more dram at ic signals such as a swaying of the upper body, especially when walking, and an increasingly deteriorating mental status. Early mental symptoms may include confusion, disorientation, vivid hallucinations, and drowsiness. Soldiers may appear to be withdrawn or demonstrate behavior generally associated with fatigue or anxiety. Like HAPE, coma or death wi 11 occur without treatment.

OTHER MOUNTAIN-RELATED ILLNESSES

1-40. Other illnesses and effects related to the mountain environment and higher elevations are:

- Subacute mountain sickness. Subacute mountain sickness occurs in some soldiers during prolonged deployments (weeks/months) to elevations above 3,600 meters (12,000 feet). Symptoms include sleep disturbance, loss of appetite, weight loss, and fatigue. This condition reflects a failureto acclimatize adequately.
- Carbon monoxide poisoning. Carbon monoxide poisoning is caused by the inefficient fuel combustion resulting from the low oxygen content of air and higher usage of stoves, combustion heaters, and engines in enclosed, poorly ventilated spaces.

Sleep disturbances. High altitude has significant harmful effects on sleep. The most prominent effects are frequent periods cf apnea (temporary suspension of respiration) and fragmented sleep Sleep disturbances may last for weeks at elevations less than 5.400 meters (18.000 feet) and may never stop at higher elevations. These effects have even been reported as low as 1,500 meters (5,000 feet).

Poor wound healing. Poor wound healing resulting from lowered immune functions may occur at higher elevations. Injuries resulting from burns, cuts, or other sources may require descent for effective treatment and healing.

ACCLIMATIZATION

1-41. Altitude acclimatization involves physiological changes that permit the body to adapt to the effects of low oxygen saturation in the air. It allows soldiers to achieve the maximum physical work performance possible for the altitude to which they are acclimatized. Once acquired, acclimatization is maintained as long as the soldier remains at that altitude, but is lost upon returning to lower elevations. Acclimatization to one altitude does not prevent altitude illnesses from occurring if ascent to higher altitudes is too rapid.

1-42. Getting used to living and working at higher altitudes requires acclimatization. Figure 1-8 shows the four factors that affect acclimatization in mountainous terrain. These factors are similar to those a scuba diver must consider, and the consequences of an error can be just as severe. In particular. high altitude dimbing must be carefully paced and staged in the same way that divers must pace and stage their ascent tothe surface.

- **Altitude**
- **Rate of Ascent**
- **Duration of Stay**
- **Level of Exertion**

Figure 1-8. Factors Affecting Acclimatization

1-43. For most soldiers at high to very high altitudes. 70 to 80 percent of the respiratory component of acclimatization occurs in 7 to 10 days, 80 to 90 percent of overall acdimatization is generally accomplished by 21 to 30 days, and maximum acdimatization may take several months to years. However, some soldiers may acdimatize more rapidly than others, and a few soldiers may not acclimatize at all. There is no absolute way to identify soldiers who cannot acdimatize, except by their experience during previous altitude exposures.

1-44. Commanders must be aware that highly fit, motivated individuals may go too high too fast and become victims of AMS. HAPE, or HACE. Slow and easy dimbing. limited activity, and long rest periods are critical to altitude acclimatization. Leaves that involve soldiers descending to lower altitudes and then returning should be limited. Acdimalization may be accomplished by either a staged or graded ascent. A combination of the two is the safest and most effective method for prevention of high altitude illnesses.

- Staged Ascent. A staged ascent requires soldiers to ascend to a moderate altitude and remain there for 3 days or more to acdimatize before ascending higher (the longer theduration, the more effedive and thorough the aoclimatization to that altitude). When possible, soldiers should make several slops for staging during ascent to allow a greater degree of acclimatization.

Graded Ascent. A graded ascent limits the daily altitude gain to allow partial acclimatization. The altitude at which soldiers sleep is the critical element in this regard. Having soldiers spend two nights at 2,700 meters (9,000 feet) and limiting the sleeping altitude to no more than 300 meters per day (1,000 feet) above the previous night's sleeping altitude will significantly reducethe incidence of altitude sickness,

1-45. In situations where there is insufficient time for a staged or graded ascent, commanders may consider using the drug acetazolamide to help accelerate acdimatization; however, commanders must ensure soldiers are acclimatized before they are committed to combat. When used appropriately, it will prevent symptoms of AMS in nearly all soldiers and reduce symptoms in most others. It has also been found to improve sleep quality at high altitudes. However, commanders should consult physidans trained in high-altitude or wilderness medidne concerning doses, side effects, and screening of individuals who may be allergic. Asa non-pharmacological method, high carbohydrate diets (whole grains, vegetables, peas and beans, potatoes, fruits, honey, and refined sugar) are effedive in aiding acdimatization.

COLD

1-46. After illnesses related to not being acdimatized. cold injuries, both freezing and nonfreezing, are generally the greatest threat. Temperature and humidity decrease with increasing altitude. Reviewing cold weather injury prevention, training in shelter construction, dressing in layers, and using the buddy system are critical and may preclude large numbers of debilitating injuries. Figure 1-9 lists the cold and snow injuries most common to mountain operations. See FM 3-97.11 and FM 4-25.11 for information regarding causes, symptoms, treatment, and prevention.

- **Frostbite (freezing)**
- **Hypothermia (nonfreezing)**
- **Trench/immersion Foot (nonfreezing)**
- **Snow Blindness**

Figure 1-9. Common Cold Weather Injuries

1-47. Altitude sickness and cold injuries can occur simultaneously, with signs and symptoms being confused with each other. Coughing, stumbling individuals should be immediately evacuated to medical support at lower levels to determine their medical condition. Likewise, soldiers in extreme pain from cold injurieswhodo not respond to normal pain medications, require evacuation. Without constant vigilance, cold injuries may significantly limit the number of deployable troops and drastically reduce combat power. However, with command emphasis and proper equipment, dothing, and training, all cold-weather injuries are preventable.

SECTION III - EFFECTS ON EQUIPMENT

- Operator/Maintenance Personnel
- Line-of-Sight
- Range
- Thermal Contrast
- Ballistics and Trajectory
- Target Detection and Acquisition
- First Round Hit Capability
- Camouflage and Concealment/Noise
- Mobility
- Wear and Maintenance
- Aerodynamics and Lift
- Functioning and Reliability
- Positioning/Site Selection

Figure 1-10. Weapons and Equipment Factors Affected by the Environment

1-48. No manual can cover the effects of terrain and weather on every weapon and item of equipment within the Army inventory. Although not all-encompassing, the list at Figure 1-10 contains factors that commanders should take into account when considering the effect the mountainous environment may have on their weapons and equipment. Of these, the most important factor is the combined effects of the environment on the soldier and his subsequent ability to operate and maintain his weapons and equipment. Increasingly sophisticated equipment requires soldiers that are mentally atert and physically capable. Failure to consider this important factor often results in severe injury. lowered weapons and equipment performance, and mission failure. The information provided within this manual, combined with the information found in weapon-specific field manuals (FMs) and technical manuals (TMs), provides the information necessary to know how to modify tactics, techniques, and procedurestowin on themountain battlefield

GENERAL EFFECTS

1-49. In a mountainous environment, the speed and occurrence of wind generally increase with elevation, and the effects of wind increase with range (depending on the speed and direction). Due to these factors, soldiers must be taught the effects of wind on ballistics and how to compensate for them. In cold weather, firing weapons often creates ice fog trails. These ice fog trails obscure vision and, at thesametime, allow the enemy to more easily discern the location of primary positions and the overall structure of a unit's defense. This situation increases the importance of alternate and supplementary firing positions,

1-50. Range estimation in mountainous terrain is difficult. Depending upon the type of terrain in the mountains, soldiers may either over- or underestimate range. Soldiers observing over smooth terrain, such as sand, water, or snow, generally underestimate ranges. This results in attempting to engage targets beyond the maximum effective ranges of their weapon systems. Looking downhill, targets appear to be farther away and looking uphill, they appear to be closer. This illusion, combined with the effects of gravity, causes the soldier shooting downhill to fire high, while it has the opposite effect on soldiers shooting uphill.

1-51. Higher elevations generally afford increased observation but low-hanging clouds and fog may decrease visibility, and the rugged nature of mountain terrain may produce significant dead space at mid-ranges. These effects mean that more observation posts are necessary to cover a given frontage in mountainous terrain than in non-mountainous terrain. They also require the routine designation of supplementary firing positions for direct fire weapons. Rugged terrain also makes ammunition resupply more difficult and increases the need to enforce strict fire control and discipline. Finally, the rugged environment creates compartmented areas that may preclude mutual support and reduce supporting distances.

SMALL ARMS

1-52 In rocky mountainous terrain, the effectiveness of small arms fire increases by the splintering and ricocheting when a bullet strikesa rock. M203 and M K-19 grenade launchers are useful for covering close-in dead space in mountainous terrain. Hand grenades are also effective Although it may seem intuitive, soldiers must still be cautioned against throwing grenades uphill where they are likely to roll back before detonation. Grenades (as well as other explosive munitions) lose much of their effectiveness when detonated under snow, and soldiers should be warned that hand grenades may freeze to wet gloves.

1-53. As elevation increases. air pressure and air density decrease. At higher elevations, a round is more efficient and strikes a target higher, due to reduced drag. This effect does not significantly influence the marksmanship performance of most soldiers, however, designated marksmen and snipers should re-zero their weapons after ascending to higher elevations. (See FM 3-25.9 and FM 3-23,10 for further information on ballistics and weather effects on small arms.)

MACHINE GUNS

1-54. Machine guns provide long-range fire when visibility is good. However, grazing fire can rarely be achieved in mountains because of the radical changes in elevation. When grazing fire can be obtained, the ranges are normally short. More often, plunging fire is the result (see Figure 1-11 and FM 3-21.7). In mountainous terrain, situations that prevent indirect fire support from protecting advancing forces may arise. When

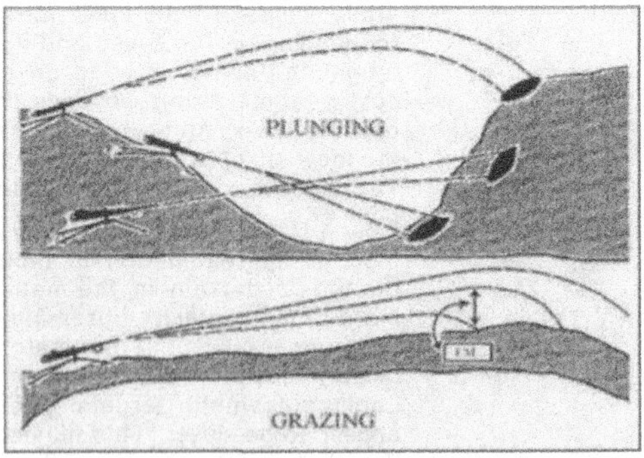

Figure 1-11. Classes of Fire with Respect to the Ground

these occur, the effects of machine-guns and other direct fire weapons must be concentrated to provide adequate supporting fires for maneuvering elements. Again, supplementary positions should be routinely prepared to cover different avenues of approach and dead space.

ANTITANK WEAPONS

1-55. The AT4 is a lightweight antitank weapon ideally suited for the mountainous environment and for direct fire against enemy weapon em placemen ts. Anti-tank guided missiles (ATGMs), such as the Javelin and the tube-launched, optically tracked, wire-guided, heavy antitank missile system (TOW), tend to hinder dismounted operations because of their bulk and weight. In very restrictive mountainous terrain, the lack of armored avenues of approach and suitable targets may limit their utility. If an armored or mechanized threat is present, TOWs are best used in long-range, antiarmor ambushes, while the shorter-range Javelin, with its fire-and-forget technology, is best used from restrictive terrain nearer the kill zone. However, their guidance systems may operate stiffly and sluggishly in extreme cold weather.

SECTION IV - RECONNAISSANCE AND SURVEILLANCE

RECONNAISSANCE

1-56. During operations in a mountainous environment, reconnaissance is as applicable to the maneuver of armies and corps as it is to tactical operations. Limited routes, adverse terrain, and rapidly changing weather significantly increase the importance of reconnaissance operations to focus fires and maneuver. Failure to conduct effective reconnaissance will result in units being asked to achieve the impossible or in missed opportunities for decisive action.

1-57. As in all environments, reconnaissance operations in a mountainous area must be layered and complementary in order to overcome enemy attempts to deny critical information to the friendly commander. In order to gather critical and timely information required by the commander, the activities of reconnaissance assets must be closely coordinated. Strategic reconnaissance platforms set the stage by identifying key terrain, as well as the general disposition and composition of enemy forces. Operational level commanders compare the information provided by strategic assets with their own requirements and employ reconnaissance assets to fill in the gaps that have not been answered by strategic systems and achieve the level of detail they require.

1-53. At the beginning of a campaign in a mountainous environment, reconnaissance requirements will be answered by aerial or overhead platforms, such as satellites, joint surveillance, target attack radar systems (JSTARSs), U2 aircraft, and unmanned aerial vehicles (UAVs). In a mountain AO, it may often be necessary to commit ground reconnaissance assets in support of strategic and operational information requirements. Conversely, strategic and operational reconnaissance systems may be employed to identify or confirm the feasibility of employing ground reconnaissance assets. Special reconnaissance (SR) and long-range surveillance (LRS) teams may be inserted to gather information thai cannot be collected by overhead systems, or to verify data that has already been collected. In this instance, satellite imagery is used to analyze a specific area for insertion for the team. The potential hide positions for the teams are identified using imagery and. terrain and weather permitting, verified by UAVs. See FM 3-100.55 for detailed information on combined arms reconnaissance.

1-59. In harsh mountain terrain, ground reconnaissance operations are often conducted dismounted. Commanders must assess the slower rate of ground reconnaissance elements to determine its impact on theentire reconnaissance and collection process. They must develop plans that account for this slower rate and initiate reconnaissance as early as possible to provide additional time for movement. Commanders may also need to allocate more forces, including combat forces, to conduct reconnaissance, reconnaissance in force missions, or limited objective attacks to gain needed intelligence. Based upon mission, enemy, terrain and weather, troops and support available, time available, civil considerations (METT-TC), commanders may need to prioritize collection assets, accept risk, and continue with less information from their initial reconnaissance efforts. In these cases, they must use formations and schemes of maneuver that provide maximum security and flexibility, to include robust security formations, and allow for the development of the situation once in contact.

1-60. Although reconnaissance patrols should normally use the heights to observe the enemy, it may be necessary to send small reconnaissance teams into valleys or along the low ground to gain suitable vantage points or physically examine routes that will be used by mechanized or motorized forces. In mountainous environments, reconnaissance elements are often tasked to determine:

- The enemy's primary and alternate lines of communication.
- Locations and directions from which the enemy can attack or counterattack.
- Heights that allow the enemy to observe the various sectors of terrain.
- Suitable observation posts for forward observers.
- Portions of the route that provide covert movement.
- Level of mountaineering skill required to negotiate routes (dismounted mobility classification) and sections of the route that require mountaineering installations.
- Suitability of routes for sustained combat service support (CSS) operations.
- Trails, routes, and bridges that can support or can be improved by engineers in order to move mechanized elements into areas previously thought to be impassable.
- Bypass routes
- Potential airborne and air assault drop^pick-up zones and aircraft landing areas.

RECONNAISSANCE IN FORCE

1-61. The compart merited geography and inherent mobility restrictions of mountainous terrain pose significant risk for reconnaissance in force operations. Since the terrain normally allows enemy units to defend along a much broader front with fewer forces, a reconnaissance in force may be conducted as a series of smaller attacks to determine the enemy situation at selected points. Commanders should carefully consider mobility restrictions that may affect plans for withdrawal or exploitation. Commanders should also position small reconnaissance elements or employ surveillance systems throughout the threat area of operations to gauge the enemy's reaction to friendly reconnaissance in force operations and alert the force to possible enemy counterattacks. In the mountains, the risk of having at least a portion of the force cut off and isolated is extremely high. Mobile reserves and preplanned fires must be available to reduce the risk, decrease the vulnerability of the force, and exploit any success as it develops.

ENGINEER RECONNAISSANCE

1-62. Engineer reconnaissance assumes greater significance in a mountainous environment in order to ensure supporting engineers are properly task organized with specialized equipment for quickly overcoming natural and reinforcing obstacles. Engineer reconnaissance teams assess the resources required for clearing obstacles on precipitous slopes, constructing crossing sites at fast-moving streams and rivers, improving and repairing roads, erecting fortifications, and establishing barriers during the conduct of defensive operations. Since the restrictive terrain promotes the widespread employment of point obstacles, engineer elements should be integrated into all mountain reconnaissance operations,

1-63. In some regions, maps may be unsuitable for tactical planning due to inaccuracies, limited detail, and inadequate coverage. En these areas, engineer reconnaissance should precede, but not delay operalions. Because rugged mountain terrain makes ground reconnaissance time-consuming and dangerous, a combination of ground and aerial or overhead platforms should be used for the engineer reconnaissance effort. Data on the terrain, vegetation, and soil composition, combined with aerial photographs and multispectral imagery, allows engineer terrain intelligence teams to provide detailed information that may be unavailable from other sources.

AERIAL AND OVERHEAD RECONNAISSANCE

1-64. During all but the most adverse weather conditions, aerial or overhead reconnaissance may be the best means to gather information and cover large areas that are difficult for ground units to traverse or observe. Airborne standoff intelligence collection devices, such as side-looking radar, provide excellent terrain and target isolation imagery. Missions must be planned to ensure that critical areas are not masked by terrain or other environmental conditions. Additionally, aerial or overhead photographs may compensate for inadequate maps and provide the level of detai l needed to plan operations. I nfrared imagery and camouflage detection film can be used to determine precise locations of enemy positions, even at night. Furthermore, AH-64 and

OH-58D helicopters can provide commanders with critical day or night video reconnaissance, utilizing television or lorward-lcoking infrared.

1-65. Terrain may significantly impact the employment of overhead reconnaissance platforms using radar systems to detect manmade objects. These systems may find themselves adversely impacted by the masking effect that occurs when the mountain terrain blocks the radar beam. Thus, the radar coverage may not extend across the reverse slope of a steep ridge or a valley floor. Attempts to reposition the overhead platform to a point where it can "see" the masked area may merely result in masking occurring elsewhere. This limitation does not preclude using such systems; however, the commander should employ manned or unmanned aerial reconnaissance when available, in conjunction with overhead reconnaissance platforms in order ta minimize these occurrences. The subsequent use of ground reconnaissance assets to verify the data that can be gathered by overhead and electro-optical platforms wi 11 ensure that commanders do not fall prey to del i berate enemy deception efforts that capitalize on the limited capabilities of some types of overhead platforms in this environment.

SURVEILLANCE

1-66. I n the mountains, surveillance of vulnerable flanks and gaps between units is accomplished primarily through well-positioned observation posts (OPs). These OPs are normally inserted by helicopter and manned by small elements equipped with sensors, enhanced electro-optical devices, and appropriate communications. Commanders must develop adequate plans that address not only their insertion, but their continued support and ultimate extraction. The considerations of METT-TC may dictate that commanders provide more personnel and assets than other types of terrain to adequately conduct surveillance missions. Commanders must also ensure that surveillance operations are fully integrated with reconnaissance efforts in order to provide a3dequate coverage of the AO,

1-67. Long-range surveillance units (LRSUs) and snipers trained in mountain operations also contribute to surveillance missions and benefit from the restrictive terrain and excellenl line-of-sight. Overhead platforms and air cavalry may also be used for surveillance missions of limited duration. However, weather may impede air operations, decrease visibility for both air and ground elements, and reduce the ability of ground surveillance elements to remain hidden for prolonged periods without adequate logistical support. As with overhead reconnaissance, terrain may mask overhead surveillance platforms.

Chapter 2

Command and Control

In the mountains, major axes of advance are limited to accessible valleys and often separated by restrictive terrain. The compart merited nature of the terrain makes it difficult to switch the effort from one axis to another or to offer mutual support between axes. The battle to control the major lines of communications of Level I develops on the ridges and heights of Level II. In turn, the occupation of the dominating heights in Level II may leave a force assailable from the restrictive terrain of Level III. Each operational terrain level influences the application of tactics, techniques, and procedures n ecessar y for su ccessf u l operations.

CONTENTS

In mountainous terrain, it is usually difficult to oonduct a coordinated battle. Engagements tend to be isolated, march columns of even small elements extremely long, and mutual support difficult to accomplish. Command and control of all available assets is best achieved if command posts are well forward. However, the mountainous environment decreases the commander's mobility. Therefore, commanders must be able to develop a clear vision of how the battle will unfold, correctly anticipate the decisive points on the battlefield, and position themselves at these critical points.

The success of a unit conducting mountain operations depends on how well leaders control their units. Control is limited largely to a well-thought-out plan and thorough preparation. Boundaries require careful planning in mountain operations. Heights overlooking valleys should be included in the boundaries of units capable of exerting the most influence over them. These boundaries may be difficult to determine initially and may require subsequent adjustment.

During execution, leaders must be able to control direction and speed of movement, maintain proper intervals, and rapidly start, stop, or shift fire. In the mountains, soldiers focus mainly on negotiating difficult terrain. Leaders, however, must ensure that their soldiers remain alert for, understand, and follow signals and orders Although in most instances audio, visual, wire, physical signals, and messengers are used to maintain control, operations may be controlled by time as a secondary means. However, realistic timetables must be based on thorough reconnaissance and sound practical knowledge of the mountain battlefield.

Commanders must devote careful consideration to the substantial effect the mountain environment may have on systems that affect their ability to collect, process, store, and disseminate information. Computers, communications, and other sophisticated electronic equipment are usually susceptible to jars, shocks, and rough handling associated with the rugged mountain environment. They are also extremely sensitive to the severecold often associated with higher elevations. Increased precipitation and moisture may damage electronic components, and heavy amounts of rain and snow, combined with strong surface winds, may generate background electronic interference that can reduce the efficiency of intercept/direction finding antennas and ground surveillance radars. Localized storms with low sustained cloud cover reduce the effectiveness of most imagery intelligence (IMINT) platforms, to include unmanned aerial vehicles (UAVs). The collective effect of mountain weather and terrain diminishes a commander's ability to achieve shared situational understanding among his subordinates. However, increased use of human intelligence (HUMI NT), clear orders and intents, and leaders capable of exercising initiative, allow commanders to dominate the harsh environment of a mountain area of operations.

As in any environment, mountain operations pose both tactical and accident risks. However, since most units do not routinely train for or operate in the mountains, the level of uncertainty, ambiguity, and friction is often higher than in less rugged environments. Commanders must be able to identify and assess hazards that may be encountered in executing their missions, develop and implement control measures to eliminate unnecessary risk, and continuously supervise and assess to ensure measures are properly executed and remain appropriate as the situation changes. Although risk decisions are the commanders' business, staffs, subordinate leaders, and individual soldiers must also understand the risk management process and must continuously look for hazards at their level or within their area of expertise. Any risks identified (with recommended risk reduction measures) must be quickly elevated to the chain of command (see **FM** 3-100.14).

SECTION I - ASSESSMENT OF THE SITUATION

2-1, Although higher-elevation terrain is not always key, the structure of a mountain area of operations (AO) often forms a stairway of key terrain features Identification and control of dominant terrain at each operational terrain level form the basis for successful mountain maneuver. Key terrain features at higher elevations often take on added significance due to their inaccessibility and ease of defense. Tomainlain freedom of maneuver, commanders must apply combat power so that the terrain at Levels II and III can be exploited in the conduct of operations. Successful appl i cat ion of this concept requires commanders to think, plan, and maneuver vertical l y as well as horizontal ly.

2-2. Mountain operations usually focus on lines of communication, choke points, and dominating heights. Maneuver generally attempts to avoid strengths, envelop the enemy, and limit his ability to effectively use the high ground. Major difficulties are establishing boundaries, establishing and maintaining communications, providing logistics, and evacuating wounded. Throughout the plan, prepare, and execute cycle, commanders must continuously assess the vertical impact on the mission, enemy, terrain and weather, troops and support available, time available, civil considerations (METT-TC),

HISTORICAL PERSPECTIVE

Importance of Controlling Key Terrain:

Mustafa Kemal at Gallipoli (April 1915)

On 25 April 1915, the Allies launched their Gallipoli campaign. However, LTC Mustafa Kemal's understanding of the decisive importance of the hilly terrain, his grasp of the enemy's overall intent, and his own resolute leadership preserved the Ottoman defenses. His troops seized the initiative from superior forces and pushed the Allied invasion force back to its bridgehead. The result was nine months of trench warfare, followed by the Allies' withdrawal from Gallipoli.

German Fifth Army Commander General von Sanders expected a major Allied landing in the north, at Bulair. The British, however, were conducting a feint there; two ANZAC divisions were already landing in the south at Ari Burnu (now known as "ANZAC cove") as the main effort. The landing beaches here were hemmed by precipitous cliffs culminating in the high ground of the Sari Bair ridge, a fact of great importance to the defense. Only one Ottoman infantry company was guarding the area. Although prewar plans had established contingencies for using 19th ID, Kemal, the division commander, had received no word from his superiors regarding the developing scenario. Nevertheless, understanding that a major Allied landing could easily split the peninsula, he decided that time was critical and set off for Ari Burnu without waiting for his senior commander's approval. In his march toward Ari Burnu that morning, he recognized that the hilly terrain in general and the Sari Bair ridge in particular were of vital strategic importance: if the enemy captured this high ground they would be in an excellent position to cut the peninsula in half.

Kemal now engaged the enemy. He impressed upon his men the importance of controlling the hilltops at all costs, issuing his famous order: "I am not ordering you to attack. I am ordering you to die. In the time it takes us to die, other forces and commanders can come and take our place." Despite being outnumbered three-to-one, the Turkish counterattack stabilized their position and prevented the Allies from capturing the Sari Bair ridge. Nightfall brought about a lull in the fighting. There was some sniping and a few local encounters on 26 April, and on 27 April Kemal finally received major reinforcements. The front stabilized and the opposing armies settled down into trench warfare. On 16 January 1926, the Allies admitted defeat and withdrew.

The 19th ID's counterattack, which prevented the ANZAC from establishing themselves on the Sari Bair ridge, may well have decided the outcome of the entire Gallipoli campaign. Despite his lack of situational knowledge, Kemal instinctively understood the enemy's intent and, recognizing the importance of controlling the hilltops and ridgelines, was committed to concentrating his combat power to seize and hold this key terrain.

Compiled from "The Rock of Gallipoli," *Studies of Battle Command*, George W. Gawrych

MISSION

2-3. Mission analysis must include the spatial and vertical characteristics of the AO. Although defeating the enemy continues to be the basic objective of tactical operations, the task of controlling specific operational terrain levels will be paramount. At brigade level and below, major tactical objectives are normally translated into tasks pertaining to seizing, retaining, or controlling specific dominating heights at either Level II or Level III. Therefore, it is imperative to identify the tasks and assets necessary to access each operational terrain level.

2-4. At any operational terrain level, defending and delaying are easier at defiles, while attacking is more difficult. Due to the compartmented terrain, units usually execute offensive missions by conducting several simultaneous

smaller-scale attacks, utilizingthefull height, width, and depth of their area of operations. Consequentiy, commanders must always consider the impact of decentralization on security.

2-5. One method of maintaining freedom of action is to seize or hotd key terrain. In the mountains, key terrain is frequently identified as terrain that is higher than that held by the enemy. Seizing this terrain often depends on long and difficult envelopments or turning movements. Therefore, the specified and implied tasks associated with mobility and sustainment, as well as command and control, must be considered in terms of their vertical difficulty.

ENEMY

2-6. An enemy will normally position forces in depth and height along likely avenues of approach. Mountain terrain facilitates wide dispersal, allowing relatively smait units to hold dominant terrain in a connected system of strong points. To prevent bypassing and envelopment attempts, the enemy may adopt a many-tiered, perimeter defense. Aside from the relative size of forces, the type of enemy units and their equipment must be compared with those of friendly forces, to include a comparison of the suitability of forces, tactics, and training. When considering the enemy's ability to operate in mountainous terrain, commanders should consider how well the enemy can accomplish the tasks and actions listed in Figure **2-1.** Again, in analyzing both enemy and friendly factors during mountain operations, the vertical, as well as the horizontal, perspective should be fully integrated into all aspects of the assessment.

- **Utilize the environment to his advantage**
- **Conduct air operations**
- **Conduct decentralized operations**
- **Utilize the terrain in Levels II and III**
- **Employ obstacles or barriers to restrict maneuverability**
- **Conduct limited-visibility operations**
- **Sustain his maneuver elements**

Figure 2-1. **Factors Affecting Assessment of the Enemy Situation**

TERRAIN AND WEATHER

2-7. As in all military operations, terrain analysis involves observation and fields of fire, cover and concealment, obstacles, key terrain, and avenues of approach (OCOKA). Terrain often influences the conduct of operations more in the mountains than on flatter terrain. The mountains form the nonlinear and vertical structure of the battlefield, and the influences of geography and climate dictate the extent to which commanders modify tactics. Examples of these difficulties are often encountered in the concentration of forces, as well as in the maintenance of command and control.

2-8. In themountains, as elsewhere, surprise is easier to achieve for the force that knows the terrain better and has the skills and equipment necessary to achieve greater mobility. The appropriate use of vertical terrain improves the element of surpriseiftheterrain has been analyzed properly to determine the best means to counter the enemy's reactions. Once the commander decides on a preliminary course of action, he should immediately initiate a detailed terrain reconnaissance,

2-9. In a mountainous environment, the terrain normally favors the defender and necessitates the conduct of limited visibility operations. Highly trained units can achieve significant tactical gains and decisive victories by exploiting limited visibility. However, limited visibility operations in mountainous terrain require precise planning, careful daylight reconnaissance, exceptionally good command and control, and a high degree of training. Imaginative and bold limited visibility operations can minimize the advantage of terrain for the defender and shift the balance of combat power to the side that can best cope with or exploit limited visibility.

OBSERVATION AND FIELDSOF FIRE

2-10. Although mountainous terrain generally permits excellent long-range observation and fields of fire, steep slopes and rugged terrain affect a soldier's ability to accurately estimate range and frequently cause large areas to be hidden from observation. The existence of sharp relief and dead space facilitates covert approaches, making surveillance difficult despite such long-range observation. Four factors that influence what can be seen and hit in mountainous terrain are listed in Figure 2-2.

1. The ability to observe and identify targets in conditions of bright sunlight
2. The ability to estimate range in clear air
3. The ability to apply wind corrections
4. The ability to shoot accurately up and down vertical slopes

Figure 2-2. Factors Affecting Observation and Fields of Fire

COVER AND CONCEALMENT

2-11.The identification and proper use of the cover and concealment provided by mountainous terrain are fundamental to all aspects of mountain operations. The ridge systems found in Level II may provide covert approaches through many areas that are hidden from observation by the vegetation and relief. The difficulties a force encounters in finding available cover and concealment along ridges are fewer than those on the peaks, especially above the timberline. Uncovered portions of an approach leave a force exposed to observation and fire for long periods. The enemy can easily detect movement in this region, leaving commanders with three primary options to improve cover and concealment:

1, Identify and exploit avenues of approach the enemy would consider unlikely, due to their difficult ascent or descent.

2, Negotiate routes during periods of limited visibility.

3, Provide overwhelming route security.

OBSTACLES

2-12, Obvious natural obstacles indude deep defiles, diffs, rivers, landslides, avalanches, crevices, and scree slopes, as well as the physical terrain of the mountains themselves. Obstades vary in their effect on different forces Commanders must evaluate the terrain from both the enemy and friendly force perspective. They must look spedfreal ly at the degree to which obstades restrid operations, and at the ability of each force to exploit the tactical opportunities that exist when obstades are employed. M an-made obstades used in conjundion with restrictive terrain are extremely effedive in the mountains; however, their construction is very costly in terms of time, materiel, transportation assets, and labor. Commanders must know the location, extent, and strength of obstades so that they can be incorporated into their scheme of maneuver.

KEY TERRAIN

2-13. Key terrain generally increases in importance with an increase in elevation and a decrease in accessibility. In the mountains, however, terrain that is higher than that held by the opposing force is often key, but only if the force is capable of fighting there. A well-prepared force capable of maneuver in rugged terrain can gain an even greater advantage over an ill-prepared enemy at higher elevation levels.

2-14. The vast majority of operations in the mountains requires that the commander designate dedsive terrain in his concept of operations to communicate its importance to his staff and subordinate commanders. In operations over mountainousterrain, the analysis of key and dedsive terrain is based on the identification of these features at each of the three operational terrain levels. There are few truly impassable areas in the mountains. The commander must recognize that what may be key terrain to one force may be an obstacle to another force. He must also recognize that properly trained combatants can use high obstruding terrain as a means to achieve dedsive vidories with comparatively small-sized combat elements.

AVENUES OF APPROACH

2-15. In mountainous terrain, there are few easily accessible avenues of approach, and they usually run along valleys, defiles, or the crests and spurs of ridges. This type of geography allows the defender to economize in difficult terrain and to concentrate on dangerous avenues of approach A typical offensive tactic is to conduct a coordinated assault with the main effort along accessible avenues of approach, and supporting efforts by one or more maneuver elements on difficult and unexpected avenues of approach. Normally, high rates of advance and heavy concentration of forces are difficult or impossible to achieve along mountainous avenues of approach. Relief features may create large areas of dead space that facilitate covert movement. Units may use difficult and unlikely avenues of approach to achieve surprise: however, these are extremely high-risk operations and are prone to failure unless forces are well trained and experienced in mountaineering techniques. In mountainous terrain, the analysis of avenues of approach should be based on a thorough reconnaissance and evaluated in terms of the factors listed in Figure 2-3 on page 2-8.

WEATHER

2-16. As discussed in Chapter 1, weather and visibility conditions in the mountainous regions of the world may create unprecedented advantages and disadvantages for combatants. To fight effectively, commanders must acquire accurate weather information about their AO. Terrain has a dominant effect on local climate and weather patterns in the mountains. Mountainous areas are subject to frequent and rapid changes of weather, including fog, strong winds, extreme heat and cold, and heavy rain or snow. Thus, many forecasts that describe weather over large areas of terrain are inherently inaccurate. Commanders must be able to develop local, terrain-based forecasts by combining available forecasts with field observations (local temperature, wind, precipitation, cloud patterns, barometric pressure, and surrounding terrain). Forecasting mountain weather from the field improves accuracy and enhances the ability to exploit opportunities offered by the weather, while minimizing its adverse effects (see Appendix B).

- **Ability to achieve surprise**
- **Vulnerability to attack from surrounding heights**
- **Ability to provide mutual support to forces on other avenues of approach**
- **Effect on rates of advance**
- **Effect on command and control**
- **Potential to accommodate deception operations**
- **Ability to support necessary CS and CSS operations**
- **Access to secure rest and halt sites**
- **Potential to fix the enemy and reduce the possibility of retreat**

Figure 2-3. Factors Affecting Analysis of Avenues of Approach

TROOPS AND SUPPORT AVAILABLE

2-17. Commanders must assess the operational and tactical implications of the restrictive environment on mobility, protection, firepower, and logistics. The complex task of arranging activities in time, space, and purpose requires commanders to fully understand the impact of elevation, weather, and visibility on the capabilities of his subordinate elements and relative combat power. Mountainous terrain and weather can greatly enhance the relative combat power of defending forces and, conversely, it can drastically reduce those of the attacking forces. For example, an infantry battalion may be inadequate to defeat a defending infantry company in the mountains. Instead, an infantry battalion may only be capable of defeating a well-positioned infantry platoon. However, commanders must carefully consider each unique situation and weigh all tangible and intangible aspects of combat power (maneuver, firepower, leadership, protection, and information) when comparing strengths and determining the forces necessary to accomplish the mission.

2-18. Commanders must also assess the proper mix of heavy and light forces that capitalizes on the unique strengths that each type of force can bring to mountain operations while minimizing their limitations. While generally complicating command and control, an appropriate mix allows commanders more flexibility in the synchronization of their operations. Additionally, the difficulty providing combat support and combat service support for mountain operations must be evaluated to determine if the proportion of support troops to combat troops is sufficient.

2-19. Prior to and throughout an operation, commanders must continually assess the effect that the rugged mountain environment and sustained combat operations has on the ability of their soldiers to accomplish the mission.

Commanders may need to slow the pace of their operation, transition to the defense for short periods, or rotate units to ensure that their soldiers are physically capable of striking effectively at decisive times and locations. Too often, commanders consider only the operational readiness (OR) rate of equipment and logistics levels when determining their overall ability to continue offensive actions. Failure to consider this intangible human aspect may result in increased loss of lives and mission failure.

2-20. Vertical operations are an integral part of mountain operations and are one means to improve the success of decisive engagements. Commanders must review the state of training of their units to ensure they are adequately prepared to maneuver and fight at various elevations. Increased requirements for aviation support require aviation units to be capable of operating in the specific mountain environment. Units must also have sufficient numbers of pathfinders and trained air assault personnel to select and mark landing zones (LZs) and prepare sling loads.

TIME AVAILABLE

2-21. In the mountains, proper timing is fundamental to creating opportunities to fight the enemy on favorable terms. Restrictive terrain, weather, the accumulation of chance errors, unexpected difficulties, and the confusion of battle increase the time necessary to assemble, deploy, move, converge, and mass combat power, effectively decreasing the amount of time available to plan and prepare. To optimize the time available, commanders must continuously evaluate the impact of reduced mobility caused by the weather and terrain. At times, commanders may need to conduct a tactical pause to facilitate the concentration of combat power at a decisive point. However, they must consider time with respect to the enemy as time available is always related to the enemy's ability to execute his own plan, prepare, and execute cycle. Figure 2-4 summarizes the time considerations that are different from or greater than those encountered on flatter terrain.

- **Adaptability of plans to the terrain and varying weather**
- **Increased time needed to conduct reconnaissance, execute movements, and synchronize events on the battlefield**
- **Significant variance in the number of hours of visibility with season and elevation**

Figure 2-4. Factors Affecting Time Available

CIVIL CONSIDERATIONS

2-22. Generally, civilian population centers will be located at the lower elevations of Level I close to sources of water and along major lines of communications. Refugees and displaced civilians may increase congestion on the already limited road and trail networks normally found in mountainous environments, further complicating maneuver and sustaining operations.

2-23. Commanders must also consider the impact of operations on the often-limited civilian resources available in the mountains. The wisdom of using local resources to lighten in-theater supply requirements must be balanced against the impact on civilians and their local economy. While the purchase of goods and services from the local economy is generally welcomed, it may serve to inflate prices and make it impossible for local civilians to purchase their own scarce and needed supplies.

2-24. In mountainous regions, commanders often encounter a populace of diverse political and ethnic orientation that may support, oppose, or be ambivalent to US operations or the presence of US forces. Depending on friendly force objectives, commanders may conduct public relations, civil affairs, humanitarian assistance, and psychological operations (PSYOP) to influence perceptions and attitudes of neutral or uncommitted parties. Even if commanders cfioose not to commit resources to enlist civilian sympathy and support, they must still adjust their operations to minimize damage and loss of life to innocent civilians.

SECTION II-LEADERSHIP

2-25. To help ease their anxiety in combat, soldiers must have confidence in their leaders. This confidence may diminish rapidly unless leaders demonstrate the ability to lead over formidable terrain and under the most difficult weather conditions. Superficial knowledge of mountain warfare and ignorance or underestimation of mountain hazards and environmental effects may result in mission failureand the unnecessary loss of soldiers' lives.

2-26. Effective leadership in mountain operations combines sound judgment with a thorough understanding of the characteristics of the mountain environment. Commanders must first develop flexible and adaptable leadership throughout the chain of command. They must then be able to understand and exploit the operational and tactical implications of the mountain environment, as well as its effects on personnel, equipment, and weapons. The keys to meeting this challenge are proper training and operational experience in the mountains. To fight effectively, leaders creatively exploit the opportunities offered by the mountain environment while minimizing the adverse effectsit can have on their operations.

2-27. Leadership rapidly becomes the primary element of combat power on the mountain battlefield. Commanders must recognize the distinctive effects created by decentralization of command, develop a depth of leadership that forms the vital link to unity of effort, and organize and direct operations that require minimum intervention. While specific situations require different leadership styles and techniques, the nature of mountain warfare generally necessitates that commanders embrace the philosophy of command and control known as mission command (see FM 6-0). This type of command and control requires subordinates to make decisions rapidly within the framework of the commander's concept and intent. Commanders must be able to accept some measure of uncertainty, delegate, and trust and encourage subordinate leaders at all levels to use initiative and act alone to achieve the desired results, particularly when the situation changes and they lose contact with higher headquarters.

SECTION III - COMMUNICATIONS

2-28. The communications means available to support operations in mountainous regions are the same as those to support operations in other regions of the world. However, rapid and reliable communications are especially difficult to achieve and maintain in mountainous areas. The mountainous environment requires electronic equipment that is light, rugged, portable and able to exploit the advantages of higher terrain. The combined effects of irregular terrain patterns, magnetic and ionospheric disturbances, cold, ice, and dampness on communications equipment increase operating, maintenance, and supply problems and require precise planning and extensive coordination.

COMBAT NET RADIO

SINGLE-CHANNEL GROUND AND AIRBORNE RADIO SYSTEMS (SINCGARS)

2-29. The Single-channel Ground and Airborne Radio System (SINCGARS) family of frequency modulation (FM) radios is good for the control of battalion and smaller-sized units operating in a mountainous environment (see FM 6-02.32 and FM 6-02.18). If available, hands-free radios, such as helmet-mounted radios, are an excellent means of communication for small unil tactics and close-in distances, particularly while negotiating rugged terrain. In colder environments, shortened battery life greatly reduces the reliability of manpacked systems that rely on constant voltage input to maintain maximum accuracy.

2-30. Si nee even a smal I unit may be spread over a large area, retransmission sites may be needed to maintain communications and increase range. These sites require extensive preparation and support to ensure the survival of personnel and the conti nued maintenance of equipment. Retransmission systems are often placed on the highest accessible terrai n to afford them the best I i ne-of-sight; however, through simple analysis, these locations are often predictable and make them more vulnerable to enemy interdiction. The importance and difficulty of maintaining adequate communications in mountainous terrain requires commanders to devote additional resources for the protection of these I i mi ted assets and operators skilled in the proper use of cover and concealment, noise and light discipline, and other operations security (OPSFC) measures.

2-31. Physical range [imitations, difficulties in establishing line-of-sight paths due to intervening terrain, and limited retransmission capabilities often make it difficult to establish a brigade and larger-sized radio net. However, commanders can, if within range, enter subordinate nets and establish a temporary net for various contingencies. In the mountains or if the mobile subscriber equipment network is not yet fully developed, commanders should consider the increased need for the improved high frequency radio (IHFR) family of amplitude modulation (AM) radios and single-channel tactical satellite communications terminals for extended distances.

SATELLITE COMMUNICATIONS (SATCOM)

- Greater freedom from siting restrictions
- Extended range, capacity, and coverage
- Mobility and rapid employment
- Extremely high circuit reliability

Figure 2-5. SATCOM Advantages

2-32. Satellite communications (SATCOM) terminals are light, small, portable ground terminals that are able to communicate in spite of rugged terrain. During operations in mountainous areas having little or no infrastructure to support command and control, satellite communications become the primary means of communications. Single channel SATCOM are currently transmitted over the ultrahigh frequency (UHF) band and readily support forces operating in the mountains, while providing worldwide tactical communications, in-theater communications, combat net radio (CNR) range extension, and linkage between elements of long-range surveillance units (LRSUs) and Army special operations foroes (ARSOF). SATCOM can network with multiple users, communicate while enroute, penetrate foliage while on the ground, and has several other advantages making it an ideal system for mountain communications (see Figure 2-5). However, limitations include restricted access, low-rate data communications, and lack of antijam capability Commanders should review FM 6-02.11 for further information on the employment of SATCOM.

COMMAND AND CONTROL (C^2) AIRCRAFT

2-33 Using C^2 aircraft can assist the commander in overcoming ground mobility restrictions and may improve communications that would otherwise limit his ability to direct the battle. In the mountains, terrain masking, while making flight routing more difficult, may provide the degree of protection needed to allow an increased use of aircraft. To avoid radar or visual acquisition and to survive, C^2 aircraft must use the same terrain flight techniques employed by other tactical aviation units. This flight method often degrades FM communications and reinforces the requirement for radio relay or retransmission sites

ANTENNAS AND GROUNDS

2-34. Directional antennas, both bidirectional and unidirectional, may be needed to increase range and maintain radio communications. Although easy to fabricate, directional antennas are less flexible and more time-consuming to set up. Positioning of all antennas is also crucial in the mountains because moving an antenna even a small distance can significantly affect reception

2-35. Antenna icing, a common occurrence at high elevations, significantly degrades communications. Ice may also make it difficult to extend or lower antennas, and the weight of ice buildup, combined with increased brittleness, may cause them to break. Antennas should have extra guy wires, supports, and anchor stakes to strengthen them to withstand heavy ice and wind loading. All large horizontal antennas should be equipped with a system of counterweights arranged to slacken before wire or poles break from the excess pressures of ice or wind. Soldiers may be able to remove wet snow and sleet that freezes to antennas by jarring their supports, or by attaching a hose to

the exhaust pipe of a vehicle and directing the hot air on the ice until it melts. However, soldiers must exercise great care to ensure that the antenna is not damaged in their attempts to dislodge the ice.

2-36. Ground rods and guy wires are often difficult to drive into rocky and frozen earth. Mountain pitons are excellent anchors for antenna guys in this type of soil. In extreme cold, ropes can be frozen to the ground and guys tied to these anchor ropes. Adequate grounding is also difficult to obtain on frozen or rocky surfaces due to high electrical resistance. Where it is possible to install a grounding rod, it should be driven into the earth as deep as possible or through the ice on frozen lakes or rivers. Grounding in rocky soil may be improved by adding salt solutions to improve electrical flow.

MOBILE SUBSCRIBER EQUIPMENT

2-37. Like FM. mobile subscriber equipment [MSE) requires a !ine-of-sight transmission path and a tactical satellite or several relay sites to overcome mountainous terrain and maintain MSE connectivity (FM 6-02.55 contains in-depth information concerning the deployment and employment of MSE).

WIRE AND FIELD PHONES

2-38. Wire is normally one of the most reliable means of communication Unfortunately, in rugged mountains and particularly during the winter months, wire is more difficult and time consuming to install, maintain, and protect. Wire may be dispensed in mountain areas by tracked or wheeled vehicle, foot, skis, snowshoes, or oversnow vehicles. As in any environment, units must periodically patrol their wire lines to ensure that they have remained camouflaged and that the enemy has not tapped into them.

2-39. Snow-covered cables and wire can cause the loss of many man-days in recovering or maintaining circuits. This can be avoided by pulling the cable from under the snow after each snowfall and letting it rest just below the surface of the snow. Trees or poles can be used to support wire. Allowances must be made for drifting snow when determining the height above ground at which to support the lines. However, when crossing roads, it is preferable to run the wire through culverts and under bridges rather than bury or raise wire overhead. In addition to ease, this technique reduces maintenance requirements associated with vehicles severing lines, particularly with higher volumes of traffic on limited road networks. If long-distance wire communications are required, the integration of radio relay systems must be considered.

2-40. Great care must be taken in handling wire and cables in extreme cold weather. Condensation and ice on connectors make connecting cables difficult and can degrade the signal path. When rubber jackets become hard, the cables must be protected from stretching and bending to prevent short circuits caused by breaks in the covering. Therefore, all tactical cable and wire should be stored in heated areas or warmed prior to installation. TC 24-20 provides more detailed information on the installation and maintenance of wire and cable.

2-41. Field phones are used in a stationary position, such as a mountain patrol base or an ambush site, although leaders must consider the weight and difficulties encountered in laying and maintaining wire in these sites of limited duration. The batteries that are used to operate field telephones and switchboards are subject to the same temperature limitations as those used to power tactical radiosets.

2-42. When used with a hands-free phone, commercially available rope with a communication wire in it is ideally suited for mountain operations. This system is lightweight and easy to manage, and provides an added measure of security during limited visibility operations. In addition to the standard uses, since it functions as both a rope and a wire, it can be used to control movement on all types of installations, and it can serve as a primary means of communication for climbing teams.

AUDIO, VISUAL, AND PHYSICAL SIGNALS

2-43. Leaders can use simple audio signals, such as voice or whistles, to locally alert and warn. Sound travels farther in mountain air. Although this effect may increase the possibility of enemy detection, interrupting terrain, wind conditions, and echoes can restrict voice and whistle commands to certain directionsand uses.

2-44. Like audio signals, visual signals such as pyrotechnics and mirrors have limited use dueto enemy detection, but may work for routine and emergency traffic at the right time and place. Blowing sand or snow, haze, fog, and other atmospheric conditions may periodically affect range and reliability.

2-45. Units should use hand and arm signals instead of the radio or voice when ever possi bl e, especi al l y wh en cl ose tothe enemy. Luminous tape on t he camouflage band, luminous marks on a compass, or flashlights may be used as signals at night over short distances. Infrared sources and receiving equipment, such as night vision goggles, aiming lights, and infrared filters for flashlights, can be used to send and receive signals at night However, an enemy outfitted with similar equipment can also detect active devices.

2-46. A tug system is a common method of signaling between members of a roped climbing team. However, tug systems are often unreliable when climbers are moving on a rope or when the distance is so great that the friction of the rope on the rock absorbs the signals. Separate tug lines can be installed in Static positions by tying a String, cord, or wire from one position to the next. Soldiers can pass signals quietly and quickly between positions by pulling on thetug line in a prearranged code.

MESSENGER

2-47. Although slew, communication by messenger is frequently the only means available to units operating in the mountains. Messengers should be trained climbers, resourceful, familiar with mountain peculiarities, and able to carry their own existence load. During the winter, advanced skiing skills may also be required. Messengers should always be dispatched in pairs. Air messenger service should be scheduled between units and integrated with the aerial resupply missions. Vehicles may also be employed to maintain messenger communications when conditions of time, terrain, and distance permit.

SECTION IV - TRAINING

2-48. Because US forces do not routinely train in a mountain environment, they must make extensive preparations to ensure individual and unit effectiveness. Ultimate success in the mountains depends largely on developing cohesive, combat-ready teams consisting of well-trained soldiers. To be successful, commanders must understand the stratification of mountain warfare, recognize the unique aspects of leadership required, and implement training programs that prepare soldiers for the rigors of mountain fighting.

2-49. In the mountains, commanders face the challenge of maintaining their units' combat effectiveness and efficiency. To meet this challenge, commanders conduct training that provides soldiers with the mountaineering skills necessary to apply combat power in a rugged mountain environment, and they develop leaders capable of applying doctrine to the distinct characteristics of mountain warfare.

- **Mountaineering skills**
- **Air assault and air movement operations**
- **Deception**
- **Stealth and infiltration**
- **Limited visibility operations**
- **Patrolling**
- **Reconnaissance**
- **Communications**
- **CS and CSS operations**

Figure 2-6. Training Areas of Emphasis

2-50. The ability to apply doctrine and tactics in mountainous environments is not as easy to develop as technical proficiency. Training, study, and garrison experimentation may provide the basis for competence. However, only through experience gained by practical application in the mountains will leaders become skilled in mountain warfare. Proficiency in the areas listed in Figure 2-6 wil l provide commanders with a degree of flexibility in the application of doctrine to a mountain area of operations.

2-51. The best combat and combat support plans cannot ensure victory unless commanders concentrate on developing a leadership climate that is derived from the human dimension of mountain warfare. The complexities of mountain combat make it extremely important to establish training programs that modify the traditional application of tactics so that units can reach their full potential. Training must simulate the tempo, scope, and uncertainty of mountain combat to create the versatility required to capitalize on the harsh environment as a force multiplier,

2-52. Competent units operate effectively in mountains and focus on the battle. Unprepared units, however, may become distracted by the environment and end up expending as much effort fighting the environment as they do fighting the enemy. Soldiers cannot be fully effective unless they have the proper clothing and equipment, and are trained to protect themselves against theeffectsof terrain and frequent and sudden changes in weather.

INITIAL TRAINING ASSESSMENT

2-53. In addition to the questions applicable to every mission, commanders must consider the following when preparing for operations in a mountainous environment:

- What kind of mountains will the unit be operating in?
  - What elevations will the unit be operating at?
  - What are the climatic and terrain conditions of the AO?
  - » Are at least two years of accurate weather reports available (see Appendix B)?
- When must the unit be ready to move?
- What training resources are needed and available?
- D Are local training areas and ranges available?
  - If not. what alternative arrangements can be made?
  - What available training areas most closely resemble the AO?
- D What special equipment does the unit require?
- D What training assistance is available?
  - Does the unit have former mountain warfare instructors, military mountaineers, or others with experience in a mountainous environment?
  - Are instructors available from outside the unit?
- P What special maintenance is required for weapons and equipment?
- What is the level of physical fitness?
- 0 What additional combat, combat support, and combat service support units are necessary to accomplish the operational missions?
  - Can specific units be identified for possiblecoordinated training?
- D Will allied and multinational troops participate?

2-54, As commanders get answers to these and other questions, they must develop training programs to bring their units to a level where they will be fully capable of operating successfully in mountainous conditions. To do this, they must establish priorities for training. The training requirements listed in Figure 2-7 are only a guide. Commanders should add, delete, and modify the tasks as necessary, depending on the specific AO, the state of readiness of their units when they begin preparations for mountain operations, and the time and facilities available (see FM 7-10).

PHYSICAL CONDITIONING

2-55. Soldiers who have lived and trained mostly at lower elevations tend to develop a sense of insecurity and fear about higher elevations - many are simply afraid of heights in general. With this in mind, leaders must plan training that accustoms soldiers to the effects of the mountain environment. Physical conditioning must be strictly enforced, since "new musde" strain associated with balance and prolonged ascents/descents quickly exhausts even the most physically fit soldiers. Even breathing becomes strenuous, given the

TRAINING REQUIREMENTS	ALL	STAFF AND LEADERS	TEAMS AND CREW MEMEBERS	SPECIALISTS
Physical Conditioning and Acclimatization	✔	✔	✔	✔
Mountain Illnesses and Injuries	✔	✔	✔	✔
Mountain Living and Survival	✔	✔	✔	✔
Mountain Navigation Techniques	✔	✔	✔	✔
Mounted and Dismounted March Planning	✔	✔	✔	✔
Communications Techniques	✔	✔	✔	✔
Weapons/Equipment Training	✔	✔	✔	✔
Additional Maintenance Requirements	✔	✔	✔	✔
Camouflage and Concealment	✔	✔	✔	✔
Obstacles		✔	✔	
Above-ground Fortifications	✔	✔	✔	✔
Level 1 Mountaineering	✔	✔	✔	✔
Level 2 and 3 Mountaineering				✔
Driver and Pilot Training				✔
Air Assault/Air Movement Operations	✔	✔	✔	✔
NBC Operations	✔	✔	✔	✔

Figure **2-7.** Mountain Preparatory Training

thinner atmosphere at higher altitudes. Therefore, training must emphasize exercises designed to strengthen leg muscles and build cardiovascular (aerobic) endurance (see FM 3-25.20). Frequent marches and dimbswith normal equipment loads enhance conditioning and familiarize soldiers with mountain walking techniques,

MOUNTAIN LIVING

2-56. Successful mountain living requires that personnel adjust to special conditions, particularly terrain and weather. To develop confidence, soldiers should train in conditions that closely resemble those they will face. Lengthy exercises test support facilities and expose soldiers to the isolation common to mountain operations. The mountain area of operations can be harsh, and training should develop soldiers who possess the necessary field craft and psychological edge to operate effectively under mountainous conditions. Although FM 4-25.10 and FM 3-25.76 do not specifically address mountain environments, much of their information applies. Regardless of the level of technical mountaineering training required, all soldiers deploying to a mountainous region should be trained in the areas listed in Figure *2-8* on page 2-**18.**

- Temperature extremes and clothing requirements
- Bivouac techniques and shelter construction
- Elevation and rarified air effects
- Hygiene, sanitation, and health hazards
- Locating and purifying water
- Food-gathering techniques

Figure 2-8. Mountain Living Training

2-57. Navigation in the mountains is made more difficult because of inaccurate mapping, magnetic attraction that affects compass accuracy, and the irregular pace of the soldiers It is easy to mistake large terrain features that are very far away for features that are much closer The increased necessity for limited-visibility operations restricts the use of terrain techniques as the primary means of determining and maintaining direction. Individuals must train to use a variety of equipment, such as a compass, an altimeter, global positioning system devices, and maps, as well as learn techniques pertaining to terrestrial navigation, terrain association, dead reckoning, resecfioning, and artillery marking (seeFM 3-25.26).

MD EQUIPMENT

2-58. Nearly every weapon or piece of equipment familiar to the soldier is affected to some degree by the mountain environment. In addition to honing skills, training must focus on the specific operational area and ways to overcome anticipated environmental impacts when using weapons and equipment.

2-59. Individual marksmanship training must emphasize the effect of wind and include practical training in wind measurement techniques and adjusted aiming points (holdoff). Practical training in range estimation techniques, combined with using laser range finders, M19 binoculars, target reference points, and range cards, helps lo overcome difficulties in range estimation.

2-60. In the conduct of their preparations, commanders should strive to increase the number of qualified snipers within their units, as they are ideal in the mountains and can be used to adversely affect enemy mobility by delivering long range precision rifle fire on selected targets. They can inflict casualties, slow enemy movement, lower morale, and add confusion to enemy operations A single sniper team in well-concealed positions, such as mountain passes, can severely impede enemy movement (see FM 3-21.20 and FM 3-91.2 for further information on sniper employment).

;E AND CONCEALMENT

2-61. The basic principles of camouflage and concealment also apply in mountain operations (see FM 3-24 3). However, certain elements must be adjusted for snow. With snow on the ground, standard camouflage nets and paint patterns are unsuitable. In areas where snow cover is above 15 percent of the background, winter camouflage nets should take the place of standard nets and temporary white paint should be used over the green portions of vehicles. In terrain with more than 85 percent snow cover, the vehicles and equipment should be solid white. However, with less than 15 percent snow cover, standard patterns should be mai ntained.

2-62. Snow provides excellent conditions for threat thermal and ultraviolet sensor detection. To counter these types of sensors, soldiers must be trained to utilize the terrain to mask themselves and their weapons and equipment from enemy detection. The mountainous terrain often limits the access routes to and from selected positions. Commanders must take appropriate measures to conceal vehicle tracks and limit movement times to periods of limited visibility. Snow presents a significant problem, making movement discipline an absolute requirement. When moving, leaders should be trained to follow the shadows along windswept drift lines as much as possible. Drivers should learn to avoid sharp turns, which are easily recognizable in the snow, and follow existing track marks where possible.

FORTIFICATIONS

2-63. Fighting and protective positions in the mountains do not differ significantly from other environments, except in areas of snow and rock (see FM 3-34.112 for more information on common survivability positions and FM 3-97.11 for positions created in snow). Digging positions in rocky ground is difficult and often impossible. If demolitions, pneumatic drills, and jackhammers are available, positions may be blasted or drilled in the rock to afford some degree of protection. More often, it will be necessary to build aboveground positions by stacking boulders, stones, and gabions. If possible, existing rock formations should be used as structural wall components.

2-64. If above-ground positions are to be used, considerable care should be taken to avoid siting them in view of any likely enemy avenues of approach. Even a two-man position is difficult to conceal if it is above the timberline. Camouflage nets and the use of background rocks are necessary to break up theoutlineof theposition and hide straight edges.

2-65. Positions should be built of the largest rocks available, wedged securely together. Extreme care should be taken that the walls are stable and not leaning or sloping downhill. An unstable wall is more of a liability than an asset, as the first Impact may cause it to collapse onto the defenders. Rocks and gabions should be stacked to systematically overlap each joint or seam to help ensure stable construction. Larger rocks or stones can be used to help bond layers of rock beneath. If possible, a layer of sandbags should be placed on the top of and around the inside of the wall. Substantial overhead cover is normally required in rocky areas. The effects of artillery bursts within and above a protective position are greatly enhanced by rock and gravel displacement or avalanche Figure 2-9 on page 2-20 shows simple examples of the right and wrong way to build these positions.

MILITARY MOUNTAINEERING

2-66. The skills required for movement are often difficult to learn and usually very perishable. Commanders must understand the application and mechanics of technical mountaineering systems needed for mobility and movement of soldiers and equipment. In the mountains, a unit may be ineffective unless it has the prerequisite technical training. However, some mountains may feature terrain that is relatively benign, requiring minimal specialized techniques. Other areas will mandate the need for more advanced mountaineering skills. One key to quickly determining the type and extent of training required is to analyze and classify the level of individual movement required according to the dismounted mobility classification table introduced in Chapter 1. Once commanders have determined the specific level and tasks required, TC 90-6-1 will provide them with detailed information on specific mountaineering techniques and equipment (described below).

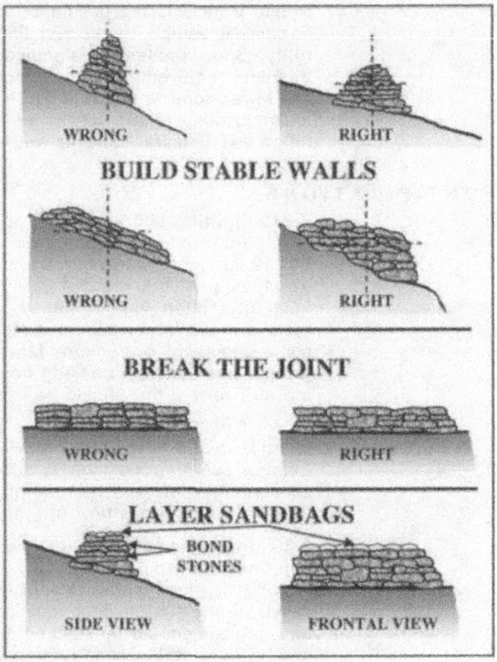

2-67. Military mountaineering training provides tactical mobility in mountainous terrain that would otherwise be inaccessible. Soldiers with specialized training who are skilled in using mountain climbing equipment and techniques can overcome the difficulties of obstructing terrain. Highly motivated soldiers who are in superior physical condition should be selected for more advanced military mountaineering training (Levels 2 and 3) conducted at appropriate facilities. Soldiers who have completed advanced mountaineering training should be used as trainers, guides, and lead climbers during collective training. They may also serve as supervisors of installation teams (see Chapter 4) and evacuation teams (see Chapter 5). Properly used, these soldiers can drastically improve mobility and have a positive impact disproportionate to their numbers. Units anticipating mountain operations should strive to achieve approximately ten percent of their force with advanced mountaineering skills.

LEVEL 1: BASIC MOUNTAINEER

2-68 The basic mountaineer, a graduate of a basic mountaineering course, should be trained in the fundamental travel and climbing skills necessary to move safely and efficiently in mountainous terrain. These soldiers should be comfortable functioning in this environment and, under the supervision of qualified mountain leaders or assault climbers, can assist in the rigging and use of all basic rope installations. On technically difficult terrain, the basic mountaineer should be capable of performing duties as the "follower" or "second" on a roped dimbing team, and should be well trained in using all basic rope systems. These soldiers may provide limited assistance to soldiers unskilled in mountaineering techniques. Particularly adept soldiers may be selected as members of special purpose teams led and supervised by advanced mountaineers. Figure 2-10 lists the minimum knowledge and skills required of basicmounlaineers

☐ Characteristics of the mountain environment (summer and winter) ☐ Mountaineering safety ☐ Use, care, and packing of individual cold weather clothing and equipment ☐ Care and use of basic mountaineering equipment ☐ Mountain bivouac techniques ☐ Mountain communications ☐ Mountain travel and walking techniques ☐ Hazard recognition and route selection ☐ Mountain navigation ☐ Basic medical evacuation	☐ Rope management and knots ☐ Natural anchors ☐ Familiarization with artificial anchors ☐ Belay and rappel techniques ☐ Use of fixed ropes (lines) ☐ Rock climbing fundamentals ☐ Rope bridges and lowering systems ☐ Individual movement on snow and ice ☐ Mountain stream crossings (to include water survival techniques) ☐ First aid for mountain illnesses and injuries

Figure 2-10. Level 1: Basic Mountaineer Tasks

2-59. In a unit training program, level 1 qualified soldiers should be identified and prepared to serve as assistant instructors to train unqualified soldiers in basic mountaineering skills. **All** high-risk training, however, must be conducted under the supervision of qualified level 2 or 3 personnel.

LEVEL 2: ASSAULT CLIMBER

2-70, Assault climbers are responsible for the rigging, inspection, use, and operation of all basic rope systems. They are trained in additional rope management skills, knot tying, belay and rappel techniques, as well as using specialized mountaineering equipment. Assault dimbers are capable of rigging complex, multipoint anchors and high-angle raising/lowering systems. Level 2 qualification is required to supervise all high-risk training associated with Level 1. At *a* minimum, assault climbers should possess the additional knowledge and skillsshown in Figure2-11 on page 2-22.

LEVEL 3: MOUNTAIN LEADER

2-71, Mountain leaders possess all the skills of the assault climber and have extensive practical experience in a variety of mountain environments in both winter and summer conditions. Level 3 mountaineers should have well-developed hazard evaluation and safe route finding skills over nil types of mountainous terrain. Mountain leaders are best qualified to advise commanders on all aspects of mountain operations, particularly the preparation and leadership required to move units over technically difficuft, hazardous, or

▯ Use specialized mountaineering equipment	▯ Movement on moderate angle snow and ice
▯ Perform multipitch climbing: - Free climbing and aid climbing - Leading on class 4 and 5 terrain	▯ Establish evacuation systems and perform high angle rescue
▯ Conduct multipitch rappelling	▯ Perform avalanche hazard evaluation and rescue techniques
▯ Establish and operate hauling systems	▯ Familiarization with movement on glaciers
▯ Establish fixed ropes with intermediate anchors	

Figure 2-11. Level 2: Assault Climber Tasks

exposed terrain. The mountain leader is the highest level of qualification and is the principle trainer for conducting mountain operations. I nstructor experience at a military mountaineering training center or as a member of a special operations forces (SOF) mountain team is critical to acquiring Level 3 qualification. Figure 2-12 outlines the additional knowledge and skills expected of mountain leaders. Depending on the specific AO, mountain leaders may need additional skills such as snowshoeing and all-terrain skiing

Figure 2-12. Level 3: Mountain Leader Tasks

DRIVER TRAINING

2-72. Driving in mountains is extremely difficult. To be successful, drivers must know their equipment's limitations and capabilities Training should center on practical exercises in mountainous terrain that gradually introduce drivers to more complex terrain and weather conditions. The exact nature of the mountainous terrain determines the training (see Figure 2-13).

- ▯ Identification and recognition of potential dangers
- ▯ Movement along steep grades combined with:
  - Narrow roads and sharp curves
  - Loose rock and gravel
  - Ice and snow (to include using tire chains for wheeled vehicles)
  - Towed loads
- ▯ Increased cold weather maintenance requirements

Figure 2-13, Driver Training

ARMY AVIATION

2-73. The mountainous environment, particularly its severe and rapidly changing weather, affects aircraft performance capabilities, accelerates crew fatigue, and influences basic flight techniques. These techniques can be acquired only through a specific training program for the particular type of mountainous terrain. Additionally, limited visibility operations in the mountains are extremely hazardous and require extensive training for those aviation units involved. Common problems associated with mountain operations become much more complex at night, even when using night vision devices. Few Army aviation units regularly train for mountain operations, so it is critical to alert them as soon as possible to fadlitate the required training to ensure safe and successful mission execution.

RECONNAISSANCE AND SURVEILLANCE

2-74. Training in reconnaissance and surveillance should focus on trafficability (route, mobility, and bridge classification), potential drop zones or landing areas, likely defensive positions, and potential infiltration routes. Infiltration and exfiltration are relatively easy in mountainous terrain and constitute a significant threat tothe maneuver elements and their support units.

TEAM DEVELOPMENT

2-75. The decentralized nature of mountain combat and the need for the exercise of a mission command philosophy of command and control involve assigning missions to i ndependenlty operating small teams that may be i sotated from thei r higher headquarters. The disruptive i nfl uences of the environment and sustained physical stress further increase the perception of i sol at ion (see FM 4-02.22). The most important factor that sustains a soldier in combat is the powerful psychological support that he receives from his primary group, such as a buddy team, squad, or platoon. H e is less l i kely to feet the stress of loneliness under the isolated conditions of mountain warfare if his primary group maintains its integrity.

2-76, The soldier's ability to survive and operate in the mountains is the basis for the self-confidence needed to feel accepted by the team. Leaders must develop small-unit cohesion down to the buddy team. Each soldier must have a buddy to share both responsibilities and rewards. The leader must not simply assign two soldiers as a buddy team, but pair soldiers whose skills and attributes complement each other. Each soldier can then learn his buddy's specialized skills adding depth to the unit if one soldier becomes disabled. Soldiers work with their buddies, as well as function as part of the larger squad team. The combined strengths of buddies enhance both unit effectiveness and combat power. FM 6-22 has more information on team development.

Chapter 3

Firepower and Protection of the Force

Employing fire support systems, which are an integral part of maneuver, is included in this chapter. This arrangement, however, does not suggest any change in the close doctrinal relationship between fires and maneuver during mountain operations.

SECTION I – FIREPOWER

FIELD ARTILLERY

3-1. The basic tactical principles for artillery remain valid in mountains, subject to the limitations imposed by terrain and weather.

MOVEMENT AND POSITIONING

3-2, Rugged terrain and reduced mobility increase the reliance on field artillery fire support. However, the employment and positioning of field artillery systems may be severely impacted by the extreme difficulty of ground mobility in mountainous terrain Self-propelled artillery is often limited to traveling on the existing road and trail networks and positioning in their immediate vicinity. Towed field artillery is usually more maneuverable; it can be brought into position with the aid of trucks, tractors, and fixed or rotary-winged aircraft. Therefore, gun crews should be

CONTENTS

proficient in equipment-rigging techniques and air assault procedures, and possess ample sling-load equipment. Field artillery emplaced by helicopter normally requires continued airlift for subsequent displacement and ammunition resupply, and often necessitates substantial engineer support,

3-3. Light field artillery may require forward displacement of gun sections by helicopter to provide forward troops the necessary support. Medium field artillery may give the longer range required, but may be limited by high-terrain crest clearance. Normally, field artillery is employed far enough to the rear to take advantage of increased angles of fall. Flat areas, such as dry riverbeds, villages and towns, and farmland, can usually accommodate firing units, however, these positions present particular problems in the mountains for the following reasons:

- Dry riverbeds are hazardous because of the danger of flash flooding.
- Towns and villages usually have adequate flat areas such as parks, schoolyards, and playing fields but they are relatively scarce and often targeted by the enemy.
- Farmland is often difficult to negotiate from spring to fall. In the winter, if the ground is frozen, farmland may provide good firing positions; however, frozen ground may cause difficulty emplacing spades, base plates, and trails.

3-4, Good artillery positions, selected for cover, flash defilade, and accessibility to road nets and landing zones (LZs). are difficult to find, and their relative scarcity makes it easier for the enemy to target probable locations. In some instances, it may be necessary to by-pass the best position for one less suitable to reduce the enemy's counterfire effects. Commanders must ensure that positions on dominant terrain provide adequate defilade. Positions on commanding terrain are preferable to low ground positions because there is-

- A reduction in the number of missions requiring high-angle fire:
- A reduced amount of dead space in the target area.
- Less exposure to small arms fire from surrounding heights.
- Less chance of being struck by rockslides or avalanches.

3-5. Seme weapons may be moved forward to provide long-range interdiction fires or. in extreme cases, direct fires to engage a road-bound enemy in mountain passes or along valley floors. Because of rugged terrain, higher angles of fire, and reduced ranges, it is generally neoessary to displace artillery more frequently than on level terrain to provide continuous support. In the mountains, commanders must often employ field artillery in a decentralized manner because of the limited space for gun positions.

ACQUISITION AND OBSERVATION

3-6. Because of high angle fire requirements, radar can be effective against enemy indirect fire systems. In many instances, terrain masking and diminished line-of-sight may degrade its effectiveness. Sites should be selected on promi nent terrain to obtain the lowest possible screening crest. However, it is often difficult to obtain a low and consistent screening crest in mountainous terrain. Too low a screening crest drives the search beam into the ground. Too high a screening crest allows the enemy to fire under the beam and avoid detection. When positioning weapons locating radars, commanders should also consider thefollowing:

- Although time-consuming, visibility diagrams are extremely useful in determining the probability of acquiring targets within the sectors of search of the radar.
- To limit search areas, radars should focus on terrain that can be occupied by artillery and mortars.
- Accurate survey control is essential because of the extreme elevation variations in mountainous terrain. Helicopters may be useful in performing survey by use of the Position Azimuth Determining System (PADS) I f possible, digital radar maps may be used to minimize the time required for height correction of the weapon system. Digital maps allow the Firefinder systems to initially locate weapon systems to within 250 meters. This allows the radar operator to make only two to three visual elevation adjustments to accurately locate the weapon system.
- Impact predict is computed at the radar's elevation, therefore, excessive errors in the impact predict can be expected.
- Firefinder radars in the same area must not face one another and radiate at the same time. This causes interference and emissions burnout, resulting in equipment failure. If radars need to face one another to accomplish the mission, commanders must coordinate to ensure that they do not radiate at the same time.
- Computing track volume may become a critical task in determining a radar's effectiveness for a proposed position (see FM 3-09,12 for computation s).
- |J Units will use more shelling reports (SHELREPs) to determine enemy firing locations.

3-7. The majority of afl field artillery fires in mountains will be observed, especially dose support and defensive fires. Unobserved fires are frequently unreliable because of poor maps and rapidly changing meteorological conditions that cause registration corrections for high angle fire to be valid for only short periods of time.

3-8. Generally, field artillery observation posts should be emplaced on the highest available ground to increase observation. Low clouds or fog may require moving them to preplanned emplacements at lower elevations. Observers must be prepared to perform assault climbing to reach the most advantageous observation site. Commanders may use aerial observers or unmanned aerial vehicles (UAVs) to detect long-range targets and complement forward observers by adjusting fires beyond terrain masks, in deep defilade, and on reverse slopes. However, in extremely high mountains aerial observers may be confined to valleys and lower altitudes due to altitude limitations on different types of aircraft,

3-9. Laser weapons demand increased emphasis on observation techniques Laser target ranging and designation systems help to overcome difficulties in range estimation by providing accurate directional distance and vertical angle information for use in locating enemy targets. However, when positioning with a laser designator, an observer should consider line-of-sight with the target, as well as cloud height. Cloud ceilings that are too low will not allow laser guided munitions enough time to lock on and maneuver to the target.

TARGETING

3-10. Because of the decentralized nature of mountain operations, targets warranting massed fires may present themselves less often than in open terrain. However, narrow defiles used as routes of supply, advance, or withdrawal by the enemy are potentially high payoff targets for interdiction fires or large massed fires. Large masses of snow or rocks above enemy positions and along main supply routes are also good targets, because they can be converted into highly destructive rockslides and avalanches that may deny the enemy the use of roads and trails, and may destroy dements in ddilade. In the mountains, suppression of enemy air defenses takes on added importance because of the increased dependence on all types of aircraft. Commanders and thdr staffs should carefully review FM 3-60. A clear understanding of the targeting methodology combined with the knowledge of the capabilities and limitations of target acquisition and attack systems in a mountain environment is crucial to the synchronization of all available combat power.

3-11. To provide accurate and timely delivery of artillery fires in mountainous terrain, commanders must take into account the foil owing:

- High angles of elevation and increased time of flight for rounds to impact.
- Targets on reverse slopes, which are more difficult to engage than targets on flat ground or rising slopes, requiring more ammunition for the same oover age.
- I ncr eased amounts of dead space that cannot be hit by artillery fires.
- I ntervening crests that require detailed map analysis.

[] When the five requirements for accurate predicted fire (target location and size, firing unit location, weapons and ammunition information, meteorological information, and computational procedures) are not achievable, registration on numerous checkpoints becomes essential because of the large variance in elevation (see FM 3-09.40 for more detailed information).

MUNITIONS

3-12 Terrain and weather also affect the use of field artillery munitions Considerations for munitions employment in the mountains are discussed below.

- Impact fuze, high explosives (HE) shells and dual-purpose improved conventional munitions (DPI CM s) are very effective on rocky ground, scattering stones and splintering rocks, which themselves become missiles. However, deep snow reduoes their bursting radius, making them approximately 40 percent less effective. The rugged nature of the terrain may afford added protection for defending forces; therefore, large quantities of HE may be required to achieve the desired effects against enemy defensive positions.
- Variable time (VT) or time fuzes should be used in deep snow conditions and are particularly effective against troops on reverse slopes. There are some older fuzes that may prematurely detonate when fired during heavy precipitation (M557 and M572 impact fuzes and M564 and M548 time fuzes).
- Smoke, DPICM, and illuminating fires are hard to adjust and maintain due to swirling, variable winds and steep mountain slopes. Smoke (a base-ejecting round) may not dispense properly if the canisters become buried in deep snow. In forested mountains, DPI CMs may get hung up in the trees. These types of munitions are generally more effective along valley floors.
- Using the artillery family of scatterable mines (FASCAM) and Copperhead is enhanced when fired into narrow defiles, valleys, and roads. FASCAM may lose their effectiveness on steep terrain and in deep sncw. Melting and shifting sncw may cause the anti-handling devices to detonate prematurely the munitions, however, very little settling normally occurs at temperatures lower than 5 degrees Fahrenheit. Remote antiarmor mine system (RAAMS) and area denial artillery munitions (ADAM) must come to rest and stabilize within 30 seconds of impact or the submunitions will not arm. and very uneven terrain may keep theADAM trip wires from deploying properly.

MORTARS

3-13. Mortars are essential during mountain operations. Their high angle of fire and high rate of fire is suited to supporting dispersed forces. They can deliver fires on reverse slopes, into dead space, and over intermediate crests, and, like field artillery, rock fragments caused by the impact of mortar rounds may cause additional casualties or damage.

3-14. The 60mm mortar is an ideal supporting weapon for mountain combat because of its portability, ease of concealment, and lightweight ammunition. The 81mm mortar provides longer range and delivers more explosives than the 60mm mortar. However, it is heavier and fewer rounds (usually no more than two per soldier) can be man-packed. The 120mm mortar may be more desirable in some situations, since they can fire either white phosphorous (WP) or HE at greater ranges than tighter mortars and have a significantly better illumination capability. However, because of the weight of these mortars and their ammunition, it may be necessary to transport fewer of them into mountainous terrain and use the remaining gun crews as ammunition bearers, or position them close to a trail network in a valley or at lower elevations. The second technique may be satisfactory if the movement of the unit can be covered and sufficient firing positions exist.

AIR SUPPORT

3-15, Air interdiction and dose air support operations can be particularly effective in mountains, since enemy mobility, like ours, is restricted by terrain. Airborne forward air controllers and close air support pilots can be used as valuable sources of information and can find and designate targets that may be masked from direct ground observation. Vehides and personnel are particularly vulnerable to effective air attack when moving along narrow mountain roads. Predsion-guided munitions, such as laser-guided bombs, can quickly destroy bridges and tunnels and, under proper conditions, cause landslides and avalanches to dose routes or collapse on both stationary and advandng enemy foroes. Moreover, air-delivered mines and long-delay bombs can be employed to seriously impede the enemy's ability to make critical route repairs. Predsion-guided munitions, as well as fuel air explosives, can also destroy or neutralize well-protected point targets, such as cave entrances and enemy forces in defilade.

3-16. Low ceilings, fog, and storms common to mountain regions may degrade air support operations. Although, global positioning system (GPS) capable aircraft and air delivered weapons can negate many of the previous limitations caused by weather. Terrain canalizes low altitude air avenues of approach, limiting ingress and egress routes and available attack options, and increasing aircraft vulnerability to enemy air defense systems. Potential targets can hide in the crevices of diffs and the niches of mountain slopes, and on gorge floors. Hence, pilots may be able to detect the enemy only at short distances, requiring them to swing around for a second run on the target and giving the enemy more time to disperse and seek better cover. Additionally, accuracy may be degraded due to the need for pilots to divert more of their attention to flying while simultaneously executing their attack.

ELECTRONIC WARFARE

3-17. The ability to use electromagnetic energy to deceive the enemy, locate his units and facilities, intercept his communications, and disrupt his command, control, and target acquisition systems remains as important in the mountains as elsewhere. Theeffectsof terrain and weather on electronic warfare (EW) systems are often *a* result of the effects on the components of those systems (particularly soldiers, communications, and aviation). Although a number of the effects are discussed in more detail elsewhere in this manual (and in applicable FMs and TMs), for ease some of the more common degrading effects of the mountainous environment on the components of electronic warfare systems are described in Figure 3-1 on page 3-8.

SECTION II - PROTECTION OF THE FORCE

AIR DEFENSE ARTILLERY

3-18. The severe mountain environment requires some modification of air defense employment techniques. Suitable positions are scarce and access roads are limited. In some instances, supporting air defense weapons may not be able to deploy to the most desirable locations. Consequently, the manportable air defense systems (MANPADS) may be the only air defense weapon capable of providing dose-in protection to maneuver elements.

3-19. Mountain terrain tends to degrade the electronic target acquisition capabilities of air defense systems. This degradation makes it more difficult for the air defense planner to locate and select position to provide adequate coverage for the force, and increases the importance of combined arms for air defense (CAFAD) and passive air defense measures (see FM 3-01.8). Individual and crew-served weapons can mass their fires against air threats. The massed use of guns in local air defense causes enemy air to increase their standoff range for surveillance and weapons delivery, and increase altitude in transiting to and from targets. These reactions may make the enemy air more vulnerable to air defense artillery (ADA).

3-20. Enemy aircraft will probably use defiles and valleys in mountainous terrain for low-altitude approaches to take advantage of terrain masking of radar. Congested roads and trails, and their junctions, may become lucrative targets for enemy air strikes. Enemy pilots may avoid early detection by using terrain-dearance or terrain-following techniques to approach a target. Rugged mountain terrain degrades air defense detection, but. at the same time, mountain ridges and peaks tend to canalize enemy aircraft. Detailed terrain analysis, coupled with predictive analysis to identify probable enemy air avenues of approach, aids in effective site selection.

EW COMPONENT	ENVIRONMENTAL FACTORS: CLOUDS	FOG	RAIN	SNOW	WIND	TEMP	TERRAIN	REMARKS
Soldiers[1]	✔	✔	✔	✔	✔	✔	✔	■ Clouds, fog, precipitation, and terrain affect visibility and observation. ■ Precipitation, temperature, and the rugged terrain affect soldier performance and ability to operate systems.
Electronics and-wire/cables[2]		✔	✔	✔		✔	✔	■ Extreme cold, combined with rugged terrain, increases fragility and breakage. ■ Precipitation and humidity affect electronic components.
Antennas[2]			✔	✔	✔		✔	■ Strong winds damage or prevent erection. ■ Precipitation and cold create ice, causing breakage (increased load and wind resistance) and reduce effectiveness. ■ Terrain affects masking and line-of-sight restrictions.
Aircraft[3]	✔	✔	✔	✔	✔	✔	✔	■ Clouds, fog, and precipitation degrade visibility and may prevent aircraft from flying under visual flight rules (VFR), precluding missions requiring aircraft landing at unimproved mountain LZs. ■ Cold and precipitation lead to icing, which impedes lift. ■ Compartmented terrain affects flight routes and target acquisition.
Vehicles			✔	✔		✔	✔	■ Rain, snow, and rugged terrain decrease mobility.
Radars/ Sensors		✔	✔	✔	✔		✔	■ Wind increases background noise, reducing efficiency. ■ Terrain affects masking and line-of-sight restrictions. ■ Fog and precipitation decrease infrared and electro-optical systems effectiveness.
Batteries						✔		■ Terrain reduces effectiveness and battery life – some systems may not even work under reduced power.

1 See Chapter 1 (Effects on Personnel)
2 See Chapter 2 (Communications)
3 See Chapter 4 (Helicopters) and the Previous Section (Air Support)

Figure 3-1. Effects of the Mountainous Environment on EW Systems

3-21 Movement to and occupation of positions in mountainous terrain require additional time. Planners must consider slope (pitch and roll), site preparation, and access route improvement prior to movement. Bradley Stinger fighting vehicle (BSFV) units often are unable to accompany small, lightly equipped maneuver elements, and may be restricted to supporting elements in more accessible areas of the battlefield. Avenger fire units can be sting-loaded by heavy lift aircraft and MANPADS airlifted into otherwise inaccessible positions. However, equipment emplaced by helicopters is resupplied and repositioned by the same means. When moving dismounted, MAIMPAO teams are limited to one missile per soldier, unless olher members of the unit are tasked to carry additional missiles.

3-22. Because of terrain masking of radars and the difficulty in establishing line-of-sigbt communications with lhe Sentinel or light and special division interim sensor (LSDIS) radar, early warning for short-range air defense (SHORAD) systems may be limited. Soldiers must maintain continuous visual observation, particularly along likely low-level air avenues of approach. Therefore, when possible. Sentinel or LSDIS radars should be emplaced on the highest accessible terrain that provides the best air picture for target detection and early warning, not necessarily peaks and summits.

ENGINEER OPERATIONS

3-23. Engineer combat support requirements increase in mountainous terrain because of the lack of adequate cover, the requirement for construction of field fortifications and obstades, and the need to breech or reduce enemy obstades. With such an enormous multitude of tasks, effective command and control of engineer assets is essential for the optimal utilization of these relatively scarce resources (see also the discussion of engineer augmentation and employment in themobility section of Chapter 4),

3-24, Digging fighting positions and creating temporary fortifications above the timberline is generally difficult because of thin soil with underlying bedrock. As described in Chapter 2, boulders and loose rocks may be used to build hasty, aboveground fortifications, Wei l-assembled positions constructed in rock are strong and offer good protection, but they require considerable time and equipment to prepare,

3-25. Engineers assist maneuver units with light equipment and tools carried in or brought into position by ground vehicles or helicopters. Bulldozers, armored combat earthmovers (ACEs). and small emplacement excavators (SEEs) can be used in some situations to help prepare positions for command bunkers and crew-served weapons. They can also be used to prepare positions off existing roads for tanks, artillery, and air defense weapons. Conventional equipment and tools are often inadequate in rocky terrain, and extensive use of demolitions may be required- In the mountains, a greater number of engineer assets will be devoted to maintaining mobility and maneuver and unit commanders should assume that available engineer support will be limited to assist them with their survivability efforts. To enhance survivability and mobility a minimum of two soldiers per maneuver platoon should be capable of using standard demolitions.

NBC PROTECTION

3-26. Terrain and weather dictate a requirement for a high degree of nuclear, biological, and chemical (NBC) defense preparedness in mountainous areas. Due to limited mobility, viable tactical positions, and limited communication abilities, friendly units must be self-sufficient in protecting themselves against NBC weapon system effects.

3-27. Wearing mission-oriented protective posture (MOPP) gear at high elevations, when possibly combined with altitude sickness, increased dehydration, and increased physical exertion, degrades performance and increases the likelihood of heat casualties. Commanders should make every effort to keep soldiers out of MOPP gear until intelligence indicators reveal that an NBC attack is imminent or it is confirmed that a hazard actually exists (see FM 3-11.4 for a discussion on vulnerability analysis). When precautions must be taken against hazards, commanders must make decisions early and allow extra time for tactical tasks. Commanders should also refer to TC 3-10 for greater detail on tactics, techniques, and procedures necessary to operate under NBC conditions.

NUCLEAR

3-28 A mountainous environment can amplify or reduce the effects of and distort lhe normal circular pattern associated with nuclear blasts. The irregular patterns reduce the accuracy of collateral damage prediction, damage estimation, and vulnerability analysis.

3-29. Air blast effects are amplified on the burst side of mountains (see Figure 3-2). Mountain walls reflect blast waves that can reinforce each other, as well as the shock front. Therefore, it is passible that both overpressure and dynamic pressure, and their duration will increase. An added danger is the creation of rockslides or avalanches A small yield nudear weapon detonated 30 kilometers or more from the friendly positions may still cause rockslides and avalanches, and easily close narrow roads and canalized passes. On the other hand, there may be little or no blast effects on the side of the mountain away from the burst.

3-30. Hills and mountains block thermal radiation, and trees and other foliage reduce it. Low douds, fog, and falling rain or snow can absorb or scatter up to 90 percent of a burst's thermal energy. During colder weather, the heavy dothing worn by soldiers in the mountains provides additional protection. However, the reflection from snow and the thin atmosphere of higher elevations may increase the effects of thermal radiation. Snow and ice melted by thermal radiation can result in flash flooding.

3-31. Frozen and rocky ground may make it difficult to construct shelters for protection from the effects of nuclear weapons. However, natural shelters such as caves, ravines, and diffs provide some protedion From nudear effects and contamination, in some instances, improvised shelters built of snow, ice, or rocks may be the only protedion avai l able. The dear mountain air extends the range of casualty-produdng thermal effeds. Within this range, however, the soldiers' added dothing reduoes casualties from these effects.

Figure 3-2. Effects of Mountains on Radiation and Blast

3-32. In mountainous regions, the deposit of radiological contamination is very erratic in speed and direction because of variable winds. Hot spots may occur far from the point of detonation, and low-intensity areas may occur very near it. Limited mobility makes radiological surveys on the ground difficult, and the difficulty of maintaining a constant flight altitude makes air surveys highly inaccurate. Additionally, melting snow contributes to the residual radiation pattern. After a nudear detonation, streams should be checked for radiation contamination before using them for drinking or bathing. As with the other effeds, the pattern of initial and induced nudear radiation may be modified by topography and the height of the burst,

BIOLOGICAL

3-33. Most biological pathogens and some toxins are killed or destroyed by the ultraviolet rays in sunlight. Above the timberline, there is little protedion from the sun; thus, the effectiveness of a biological attack may be reduced Downwind coverage may be greater because of the frequent occurrence of high winds over mountain peaks and ridges. Additionally, inversion conditions favor the downwind travel of biological agents through mountain valleys. Typically, winds flow down terrain slopes and valleys at night and up valieys and sunny slopes during the day. The effeds of mountainous terrain and rapidly changing wind conditions on the ability to predid and provide surveys of contamination for biological agents are similar to that for nudear radiation.

3-34 Temperatures and humidity also affect the survivability of biological agents. Generally, cool temperatures favor survival, and higher humidity increases the effectiveness of the agents. Extreme cold weather and snow deposited over a biologically contaminated area can lengthen the effective period of the hazard by allowing the agent to remain alive but dormant until it is disturbed or the temperature rises, if the use of biological agents is known or suspected, commanders should ensure that soldiers pay added attention to personal hygiene and consume only purified/treated water.

CHEMICAL

3-35. Wind and terrain can also cause the effectiveness of chemical agents to vary considerably. Depending on conditions, effects can be significantly enhanced or almost ineffective. High winds and rugged terrain cause chemical agent clouds to act in a manner similar to radioactive fallout. Inversions in mountain valleys may afso effectively cap an area, slowing the dissipation rate. Because of terrain and winds, accurate prediction of the downwind travel of toxic agent clouds is difficult

3-36. In mountain warfare, chemical munitions are likely to be delivered by air. The generally cooler daytime temperatures in mountainous terrain slow the evaporation process, thus allowing a potential contamination hazard to remain active longer Midday temperatures favor using persistent or blister-type agents, since nonpersistent agents dissipate too rapidly to cause any effect and unsupervised personnel are more likely to remove protective clothing for oomfort

3-37. The actions to protect against chemical agents in the mountains are not significantly different than from the requirements in less mountainous terrain. However, in extreme cold weather, survey and monitoring is often limited to the individual team mission, the FOX system may be limited to roads and trails, and the detection of vapor hazards is limited when the temperature falls below 32 degrees Fahrenheit. Decontamination may be more difficult due to freezing conditions, and the virulency period of contamination hazard for persistent agents may increase.

SMOKE AND OBSCURANTS

3-33. Smoke operations in mountainous areas are characterized by difficulties encountered due to terrain and wind Inadequate roads enhance the military value of existing roads, mountain valleys, and passes and add importance to the high ground that dominates the other terrain. Planners can use smoke and flame systems to deny the enemy observation of friendly positions, supply routes, and entrenchments, and degrade their ability to cross through tight, high passes and engage friendly forces with direct and indirect fires.

3-39. Thermally induced slope winds that occur throughout the day and night increase the difficulty of establishing and maintaining smoke operations, except in large and medium sized valleys. Wind currents, eddies, and turbulence in mountainous terrain must be continuously studied and observed, and their skillful exploitation may greatly enhance smoke operations rather than deter them. Smoke screens may be of limited use, due to enemy aerial observation, to include UAVs. and observation by enemy forces located on high ground. Smoke units may be required to operate for extended periods with limited resupply unless petroleum, oils, and lubricants (POL) supplies are emplaced in hide positions with easy access.

Chapter 4

Maneuver

The mountain environment requires the modification of tactics, techniques, and procedures. Mountains limit mobility and the use of large forces, and restrict the full use of sophisticated weapons and equipment. These limitations enable a well-trained and determined enemy to have a military effect disproportionate to his numbers and equipment. As such, mountain campaigns are normally characterized by a series of separately fought battles for the oontrol of dominating ridges and heights that overlook roads, trails, and other potential avenues of approach. Operations generally focus on smaller-unit tactics of squad, platoon, company, and battalion size. Because access to positions is normally difficult, adjacent units often cannot provide mutual support and reserves cannot rapidly deploy. Attacks in extremely rugged

CONTENTS

terrain are often dismounted, with airborne and air assaults employed to seize high ground or key terrain and to encircle or block the enemy's retreat. While the mountainous terrain is usually thought to offer the greatest advantage to the defender, the attacker can often gain success with smaller forces by effectively using deception, bold surprise actions, and key terrain.

Although mountains often increase the need to employ light forces, commanders should not be misled into believing that this environment is the sole domain of dismounted units. On the contrary, the integrated use of mounted and dismounted forces in a mountainous environment, as elsewhere, increases a commander's capabilities while reducing his limitations. However, the employment of mixed forces must be based on sound mission, enemy, terrain and weather, troops and support available, time available, civil considerations (METT-TC) analysis of the specific mountain area of operations (AO), The infantry, armor, and combined arms series of field manuals, at both battalion and brigade level, provide the capabilities and limitations for each type force, planning and safety considerations, as well as, various concepts for employment. In all cases, commanders should assign complementary missions to each type force that capitalizes on their strengths and reduces their weaknesses, and takes into consideration the differential in operations tempo. Working together on the mountain battlefield, armored and dismounted forces can offset each other's weaknesses and provide much greater lethality than any one alone.

SECTION I - MOVEMENT AND MOBILITY

4-1. To move decisively in all directions without losing momentum in a mountain area of operations requires meticulous planning and careful preparation. In a mountainous environment, numerous conditions exist that affect mobility. The force that can maintain its momentum and agility under these conditions has the best chance of winning. Reduced mobility is a primary limitation to be considered during all phases of planning for mountain operations. Rugged terrain, the time of year, the weather, and the enemy have a decisive influence on movement in the mountains. Commanders must ensure that they have sufficient time and space to deploy their forces for battle by maintaining constant security and selecting proper routes and movement techniques. Additionally, they must closely manage limited off-road areas. Tactical operations centers, artillery units, aid stations, air defense artillery, battalion trains, and other supporting units will compete for limited space in restrictive mountainousterrain.

4-2. At any elevation level, movement is generally considered to be either movement across or along terrain compartments. When moving across terrain compartments from one ridge to another, elements should use bounding overwatch. Lead elements should secure the high ground and provide overwatching fires as the rest of the element crosses the low ground. When moving along a terrain compartment, forces should move on the high ground without silhouetting t hem selves or, at a minimum, place an element there to securetheir flanks

4-3. Maneuver forces should move by stealth and exploit the cover and concealment of terrain. Using rough, unlikely routes and movement during limited visibility helps avoid enemy detection. All movements must exploit known weaknesses in enemy detection capabilities. Whenever possible, movement should be planned to coincide with other operations that divert the enemy's attention,

4-4. Because of the narrow routes sometimes encountered, especially in the higher elevations, formations may be compressed to columns or files. To reduoe vulnerability to the enemy, forces should move separated from each other on multiple and unlikely routes. When moving dismounted along unlikely routes, special teams construct fixed ropes, hauling systems, traverse systems, and other mountaineering installations to provide access to higher elevation levels and increased mobility.

4-5. The danger of surprise attack is most acute in terrain that makes deployment from the march impossible. Even with well-thought-out movement plans, maneuver elements must take both active and passive security measures at all times. Restrictive terrain facilitates templating and determining the movement of forces, making the actions of an armored force more predictable. Elements may avoid detection by using planned fires to destroy known enemy sensors and observation posts or by pfadng fires to divert the enemy's attention away from an exposed area through which the element must move. However, the placement of fires in a particular area or along a route may compromise operational security.

4-6. When the danger of rockslides or avalanches exists, the distance between elements should be increased as much as four to six times more than required on flat terrain. The more the conditions vary for each unit, the more thorough the planning must be, especially if units must reach the objective simultaneously. Often, a reserve of time must be programmed if units move on multiple routes, over unfamiliar terrain, or during limited visibility, or if they face an uncertain enemy situation.

MOUNTED MOVEMENT

WHEELED AND TRACKED VEHICLES

4-7. Generally, the mountain terrain above the valley floor severely limits movement of wheeled vehides and is too restricted for tracked vehides. Trafficable terrain tends to run along features with steep slopes on either side, making mounted movement vulnerable to vehicular ambushes and attack aircraft. Recovery vehides must always accompany mounted forces in mountainous terrain to rapidly remove disabled vehides from the limited and narrow trail network.

4-8. Tanks and other armored vehides, such as infantry fighting vehides (IFVs), are generally limited to movement in valleys and existing trail networks at lower elevations. Even at these levels, the trails may require extensive engineer work to allow tracked vehides to pass over them. Tanks,

Bradley fighting vehicles (BFVs), and cavalry fighting vehicles (CFVs) can support by fire if accessible firing positions are available; however, it will rarely be possible for them to accompany dismounted infantry in the assault. In such cases, commanders may seek to use their increased firepower to isolate the objective for the dismounted assault. If employed above Level I, armored vehicles are forced to fight in smaller numbers, yet a single tank at a critical pant may have a decisive effect. Although antitank weapons employed from higher elevations can easily penetrate the top of armored vehides, in many situations, the inability to elevate the weapon system's main gun suffidently to return fire may further increase its vulnerability.

4-9. Low atmospheric pressure considerably increases the evaporation of water in storage batteries and vehide cooling systems, and impairs cylinder breathing. Consequently, vehicles expend more fuel and lubricant, and engine power is reduced by four to six percent for every 1,000-meter (3.300-foot) i ncrease in elevation above sea level. This translates to a fuel and oil i n crease of approximately 30 to40 percent or more.

4-10. Figure 4-1 contains questions that are part of any mounted movement plan. Limited road networks and restricted off-road mobility significantly increase their importance. I n the mountains, failure to address these questions in detail may seriously jeopardize the overall mission.

HELICOPTERS

4-11, Utility and cargo helicopters are key to the rapid movement of soldiers and equipment in the mountains. However, any operation that depends primarily on continuous aviation support to succeed is extremely risky. H rgh eievat ions and rapidly changing and severe weather common to mountainous regions is very restrictive to aviation operations and makes availability of aviation support very unpredictable. At high altitudes, weather that appears to be stable to the ground observer may significantly affect helicopters- The effects of fog, frontal systems, winds, and storms are readily discernible. Additionally, higher altitudes may restrict aircraft lift capabilities and decrease aircraft allowable gross weight in mission profi le. Ai rcraft idng is common at high altitude and may occur suddenly. De-ice/anti-ice capabilities exist for rotor blades, however, idng may still decrease lift and. in severe cases, prevent flight altogether. Therefore, commanders must become intimately familiar with the conditions that may limit the full effectiveness of Army aviation when operating in a mountain environment (see FM 3-04.203).

- ☐ How fast can the march be conducted?
- ☐ Will there be other traffic on the route?
- ☐ Are there potential areas that offer covered, off-road positions?
- ☐ Are there any locations along the route that could be used for resupply?
- ☐ Are there alternate routes?

Figure 4-1. Mounted Movement Planning

4-12. Additionally, commanders must consider the effect of altitude on soldiers when planning air assault operations (see Chapter 1). If possible, commanders should use soldiers acdimatized at or above the elevation level planned for the air assault. Depending on the situation, it may be better to have troops walk in rather than fly them to the necessary elevation level.

4-13. Rugged, mountainous terrain complicates flight route selection and places an additional navigational load and strain on the entire crew, as they have little margin for error. Direct routes can seldom be flown without exposing aircraft to an unacceptable risk of detection and destruction by the enemy. Tactical flight routes follow valley corrida's, where it is possible to obtain cover and concealment while maintaining the highest possible terrain flight altitude. Terrain flight in the mountains may preclude using dosed formations. Multi-helicopter operations are normally flown in "loose" or "staggered trail" format ions with increased spadng between aircraft.

4-14. Terrain suitable for multiple helicopter landing zones (LZs) in mountainous regions is limited. Level areas that are suitable for mountain LZs frequently require little preparation beyond the dearance of loose material, since the ground is usually firm enough to support helicopters. Conversely, if LZs must be developed, dearing may be difficult due to the rocky ground. Stand-off space from rock wall faces must be deared and a level landing surface must be created. Demolitions may be required to dear large rocks but care must be used to prevent rockslides or avalanches started by the explosive shock. During the winter, snow must be packed to prevent whiteouts. Similarly, sandy or dusty LZs should be dampened with water to prevent brownouts.

4-15. When only single aircraft landing zones are available, in-flight spadng between helicopters must be significantly increased. Although helicopter LZs should be located on the windward side of ridges or peaks to take advantage of the more stablewinds, conceal ment from enemy observation and the mission are the most i m port ant factors in site selection in forward areas. When it is impossible for helicopters to land, personnel may rappel and light equipment may be sling-loaded into a LZ or. in some situations, lowered by rope while the helicopter hovers However, this may increase turnaround time and aircraft vulnerability Since available landing sites are often limited, the enemy can be expeded to target all likely locations. Personnel should secure terrain that dominates a landing site before using it. They must extensively suppress enemy air defense weapons during air assault or supply operations

4-16, Attack helicopters can be well suited for a mountain environment; however, commanders must be continuously mindful of weather and elevation effects on their employment. They can be the commander's most mobile maneuver forces in mountain warfare, enabling him to concentrate combat power quickly and exploit enemy weaknesses. During stable weather conditions, attack helicopters equipped with a variety of ordnance can rapidly engage targets beyond the range of other weapons or those masked by intervening crests. As discussed earlier, higher altitudes and icing conditions affect lift and subsequently armament loads. Ice can also prevent attack helicopters from firing their weapons altogether.

4-17, Employment time and fuel consumption increases because of the few direct routes. Terrain compartments provide excellent terrain masking and radar and visual acquisition avoidance, and allow for rapid movements to the flanks and rear of an isolated enemy force. However, these same compartments may limit aircraft maneuverability and necessitate smaller flight formations, which, in turn, may affect target engagement techniques. The compartmented terrain, combined with extended distances, may require engagement without the support of other combined arms. If terrain precludes placement of fuel and arming points in the forward area, turnaround time increases and on-station lime decreases. Since ground-to-air communication is often degraded by intervening terrain, in-flight operational control over extended ranges may be difficult.

4-18. Enemy motorized and mechanized forces may be slowed and canalized as they move up steep grades, down narrow valleys, and along mountain trails. These types of conditions allow attack helicopters to engage slower moving targets that have little room to maneuver or hide. However, these same conditions also make it difficult for pilots to select positions that allow line-of-sight to the target, sufficient tracking distance, acceptable standoff range, and adequate cover and concealment. Positions located high on a ridgeline may support successful target acquisition, tracking, and standoff, but create dangerous silhouettes and look-down angles (the angle from the aircraft to the target) that exceed aircraft weapon constraints. Lower positions, possibly in draws or saddles, may provide concealment to the flanks and an extensive backdrop to help conceal positions, but can decrease a pilot's ability to locate and track targets. Intervisibility lines may mask targets, and extreme terrain relief within the aircraft's optics field of view may inhibit tracking.

4-19. Remote Hellfire engagements avoid most of these problems (see Figure 4-2), but they may increase the time of flight of the missile. A remote engagement limits the number of aircraft exposed for tracking and lasing targets. When engaging with remote fires, a designating team is plaoed in a position overwatching the enemy's route of march. The remaining aircraft position themselves in covered and concealed battle positions oriented on the engagement area. If possible, these battle positions should be located below the reverse military

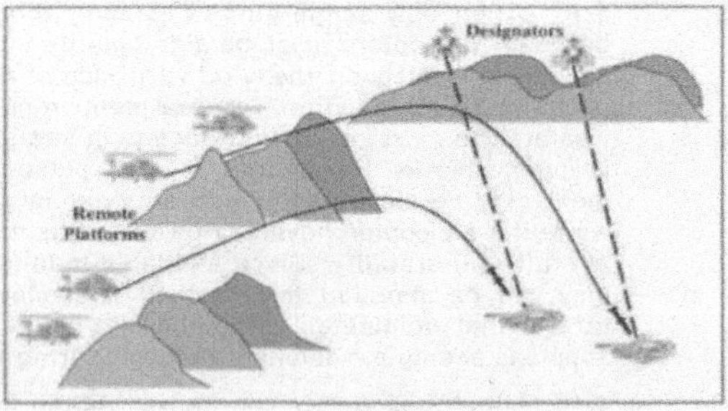

Figure 4-2. Remote Hellfire Engagements

crest or the counterslope of the ridgeline paralleling the engagement area. Aircraft in these battle positions will act as remote platforms by providing missiles to the designators without unmasking. This tactic prevents the enemy from achieving lineof sight on the firing aircraft.

4-20. The AH-64D Longbow Apache will provide an even greater killing capability for the mountain commander. It is able to detect, classify, prioritize, and engage targets with Longbow H el If ire missiles without visually acquiring the target. The commander of the AH-64D Longbow Apache utilizes the its fire control radar and mast-mounted site to target all of the vehicles in the enemy's march formation. A data transfer handover is then executed to all other AH-64D Longbow Apaches within the company. After the data transfer has been completed, the company can then engage the enemy march column without exposing the remote platforms. When employing the AH-64D Longbow Apache with radar guided Heflfire missiles, engagement times are greatly reduced and aircraft exposure to the enemy is limited. Regardless of the type of helicopter, a thorough terrain analysis and early involvement of aviation operational planners are key to successful application of Army aviation assets in mountainous terrain.

DISMOUNTED MOVEMENT

4-21. Dismounted movement is often extremely slow and arduous, and may require the skills of technical mountaineering teams to secure the advance. For example, movement in Level 11 may dictate that elements secure the high ground in Level II I. As with any type of movement, proper movement techniques and formations and conslant security to avoid unplanned enemy contact are some of the keys to successful dismounted movement.

4-22. Foot marches in the mountains are measured in time rather than distance. When making a map reconnaissance, map distance plus one-third is a good estimate of actual ground distance. One hour is added for each 300 meters of ascent or 600 meters of descent to the time required for marching a map distance. Figure 4-3 on page 4-8 shows dismounted movement calculations for an estimated 16-kilometer march on flat and mountainous terrain. Although not included in this example, commanders must also consider acclimatization, fatigue, soldiers' loads, limited visibility, and other factors that affect movement times (see FM 3-25.18 for additional factors influencing dismounted march rates).

4-23. Commanders cannot permit straggling or deviations from the selected route. Every aspect of march discipline must be rigorously enforced to keep a column closed with the knowledge that the interval between individuals depends on terrain and visibility.

4-24. To prevent an accordion effect, soldiers must allow enough distance between themselves todimb without causingthe following individual to change pace. In mountainous terrain, a slow, steady pace is preferred to more rapid movement with frequent halts. Commanders must incorporate scheduled rest halts into their movement plans based on distance, availability of covered and concealed positions, and other fadors described above. If possible, commanders should not condud rest halts during steep ascents or descents. At the start of a march, soldiers should dress lightly so that they begin slightly chilly. However, a short halt should be taken to adjust dothing and equipment after the first 15 minutes of movement. In addition, soldiers must put on special mountaineering equipment before reaching steep terrain.

Flat Terrain	Mountain Terrain
Total 16 kilometers / 4 kilometer per hour = 4 hours	Normal Time 16 kilometers / 4 kilometer per hour = 4 hours
	Ascent 600 meters / 300 meters per hour = 2 hours
	Descent 600 meters / 600 meters per hour = 1 hour
	Total 4 + 2 + 1 = 7 hours

Figure 4-3. Example Dismounted Movement Calculations

4-25. In glacial areas, the principal dangers and obstacles to movement are crevices, snow and ice avalanches. Exposure to the hazards of glaciated mountainsis increased at company-level and above, and movement should be limited to separate platoon and lower levels. When moving on glaciers, an advance element should be used- This group identifies the best routes of advance, marks the trail, and provides directions and distances to follow-on units. A marked trail is especially important during inclement weather and low visibility, and provides a route for retrograde. Commanders must carefully weigh the advantages of a marked route against the possibility of ambush and the loss of surprise.

MOBILITY

4-26. During mountain operations where limited mobility exists, it is critical that units maintain security and control of available road/transportation networks. This includes securing key bridges, fords, crossing sites, intersections, and other vulnerable choke points. These locations must be protected against enemy air. obstacle, and ground threats. However, commanders must carefully balance their available combat power between protecting their freedom of mobility/maneuver and allocating forces to critical close combat operations. Effective risk analysis and decisions are essential. Route clearance operations, patrols, traffic control points (TCPs), and other security operations aid commanders in securing routes. During offensive operations, commanders may need to commit forces to seize key terrain and routes that afford their forces greater mobility and tactical options against the enemy.

4-27. Engineer support in front of convoys and combat formations is often necessary to dear and reduce obstades, such as washouts, craters, mines, landslides, and avalanches, as well as, snow and ios in colder regions. Redudng obstades is more difficult in mountainous areas because of reduced maneuver space, lack of heavy equipment, and an increased competition for engineer support. Minefields should normally be breached, since bypassing properly sited obstades is often impossible. In the mountains, using mechanical mine plows and rollers is frequently impossible due to the lack of roads and trails, and removal of mines by hand or through demolitions is often required. Commanders must exerdse extreme caution when employi ng demolitions in the vidnity of snow and rock covered slopes because they can cause dangerous rockslides, avalanches, and secondary fragmentation FM 3-34 2 has information on breaching operations and synchronization required.

4-2S. Creating new road systems in mountainous regions is usually impractical because of the large amount of rock excavation required. Therefore, roadwork is generally limited to the existing roads and trails often requiring extensive construction, improvement, maintenance, and repair towithstand the increased military traffic and severe weather conditions. In certain mountainous areas, materials may be difficult to obtain locally and impossible to make full use of conventional heavy engineer equipment for road and bridge construction or repair. In such cases, large numbers of engineers are required and units must rely heavily on hand labor, light equipment, and demolitions.

4-29. Secondary roads and trails should be steadily improved to accommodate trucks and infantry fighting vehides, and. eventually, heavier vehicles. Their seledion depends on necessity and the speed with which the routes can be put into service. Abnormal gradients on roads may be necessary to ensure that construction keeps pace with lactical operations. Sidehill cuts are the rule, and the same contour line is followed to avoid excessive fills or bridging. Turnouts should be installed approximately every 500 meters to reduce traffic congestion on single-lane roads or trails. Drainage requirements must be considered in detail because of the effeds of abnormally steep slopes, damaging thaws, and heavy rains.

4-30. Stream and river crossing operations are difficult and must usually be accomplished by expedient means. Bridging operations in mountainous terrain are normally limited to spanning short gaps and reinfordng existing bridges by using prefabricated materials and fixed spans from floati ng bridge equipment. However, standard design or improvised suspension bridges may still be needed for longer spans. Because exist i ng bridges may have low vehide load classifications, standard fixed tactical bridges and bridging materials should be on hand to quickly ranforce or repl ace them. I n extremely rough terrai n, cableways and tramways may be construded to move l ighl loads and personnel across gorges, and up and down steep slopes.

COUNTERMOBI LITY

4-31. Obstades become more important because of the compartmented terrain and already limited road and trail networks. It is easy to create effective obstructions in mountains by cratering roads, fully or partially destroying bridges, or inducing rockslides and avalanches. Units can use antitank minefields effectively to canalize the enemy, deny terrain, or support defensive positions. Commanders should remember that dearing or redudng these same obstades may be extremely difficult and a hindrance to future operations. Using reserve and situational obstades. lanes and gaps, and plans to rapidly reduce friendly obstades must be an integral part of all defensive operations. Commanders must also consider the enemy's ability to create similar obstades and minefields when developing courses of action that hinge on speed of movement or a particular avenue of approach.

4-32. Reinforcing obstacles can be used effectively with the natural ruggedness of mountains to deny the enemy terrain and to delay and impede his movement. As in all environments, the engineer and maneuver force commander must site obstacles based on terrain and the availability of weapon systems.

4-33. Antitank mines are laid along the comparatively narrow approaches suitable for mounted attacks. Flash floods and excessive runoff may dislodge mines from their original location; however, they normally remain armed. Family of scatterable mines (FASCAM), particularly artillery-delivered and helicopter-delivered mines, increases the flexibility of the maneuver unit commander, reduces the engineer effort, and is a valuable resource in protecting rear areas from enemy envelopment and breakthroughs. Using FASCAM should be weighed against thetime in delivery, displacement of the artillery, and the additional logistics burden that may be involved

ENGINEER AUGMENTATION AND EMPLOYMENT

4-34. Because mountain terrain requires small-unit decentralized operations, an engineer platoon or company should be allocated to each maneuver battalion, light or heavy. Allocation in this manner may leave division and brigaderear areas short of engineer support.

4-35. An additional corps engineer battalion (wheeled) and an engineer light equipment company may be needed to augment an infantry division. Platoons from the engineer light equipment company may be tasked to assist divisional platoons with the engi neer effort in each maneuver battalion area. The corps combat engineer battalion (wheeled) provides heavy equipment and dump trucksrequired to support road improvement and maintenance in division and brigade rear areas. Also, this corps combat engineer battalion (wheeled) can accomplish such tasks as constructing or reinforcing bridges. To operate efficiently, additional items, such as compressors, jackhammers. power drills, chain saws, and bulldozers, may be necessary, as well as large amounts of explosives and obstacle materials.

SPECIAL PURPOSE TEAMS

4-36. On steep, exposed, or technically difficult terrain, soldiers with advanced mountaineering skills may be required to maintain or improve mobility. Advanced climbers may deploy ahead of maneuver forces during limited visibility or inclement weather to erect aids that will help maneuver elements move in difficult terrain. They may also be committed to lead forces or to operate independently to strike the enemy suddenly over unlikely routes and to occupy certain key heights that can be defended easily because of their position. The specific employment of special purpose teams is based on the mission, tasks, and requirements of the commander.

4-37. Commanders must analyze operational terrain levels and identify the mobility requirements necessary to obtain and maintain freedom of both tactical maneuver and operational movement (see Figure 4-4). It is critical that special purpose teams are properly organized before a mission begins. Once movement is underway, unplanned deviations have little chance of success. Bypassing obstacles in mountainous terrain is almost always difficult or impossible. In many instances, the best available bypass will channel friendly forces into enemy kill zones or ambushes.

4-38. To enable a force to move personnel, equipment, and supplies on the mountain battlefield with limited delays due to terrain, visibility, or obstacles, commanders should organize soldiers with advanced mountaineering training as guides, lead climbing teams, installation teams, and evacuation teams (evacuation teams are covered in the combat health support portion of Chapter 5).

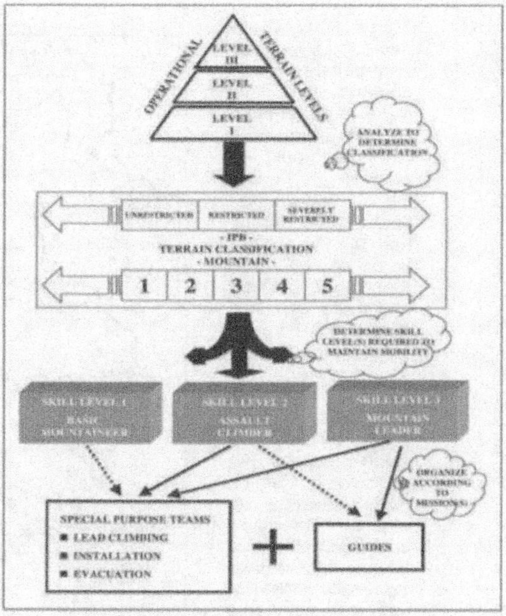

Figure 4-4. Relationship of Terrain to Skills and Special Purpose Teams

HISTORICAL PERSPECTIVE

The 10th Mountain Division and Riva Ridge (February 1945)

After attempts to capture Bologna in Italy's Po Valley during the winter of 1944-1945 failed, Allied commanders focused on the possibility of wide sweeping movements aimed at encircling Bologna and all the German armies in the region. The plan called for using the 10th Mountain Division to capture a series of mountain peaks and ridges dominating a 10-mile section of Highway 64, one of two main routes leading from Pistoia through the northern Apennines to Bologna. This would provide the Allies with a better position before starting the spring offensive, since seizure of this terrain would result not only in the Germans' being unable to protect their vital lines of communications, but also in the Americans' being able to observe German activity almost all the way to the Po Valley, approximately 40 miles away. The Division's baptism-by-fire was to be in a region dominated by two ridges whose highest peaks rose 3,000 to 5,000 feet. One of these ridges, Pizzo di Campiano-Monte Mancinello, became known as Riva Ridge (see Figure 4-5).

These heights appeared to be impregnable, as it was doubtful that any force large enough to overwhelm the Germans could be massed unobserved for an assault. The Germans had all the advantages of the commanding heights, and there was little cover for troops crossing the barren, snow-covered ground. It was clear that Riva Ridge would have to be cleared before the decisive attack could advance up Mount Belvedere and along the ridge towards Monte della Torraccia. Therefore, the plan was for the mountain troops to climb the 1500-foot cliff and surprise the Germans, who would not be expecting the attack up the face of the cliff.

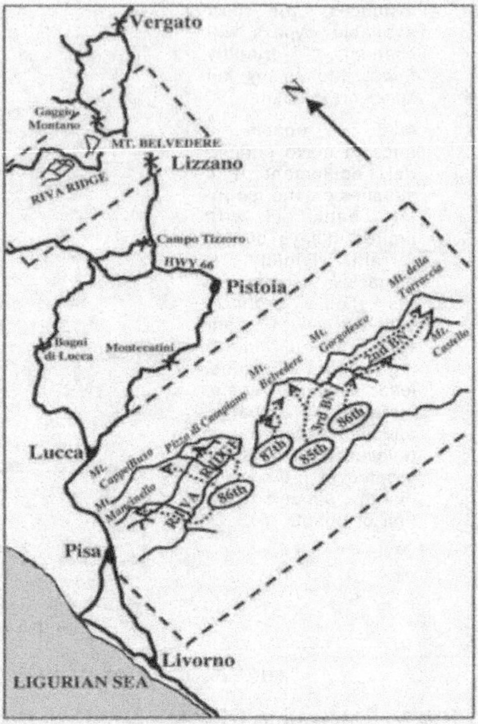

Figure 4-5. Riva Ridge

The 86th Mountain Infantry Regiment's 1st Battalion and one company from its 2nd Battalion were assigned to negotiate the cliff and capture Riva Ridge, thereby setting the conditions for other assault elements to achieve their objectives. Select teams of climbers assembled their ropes, pitons, and other gear. The teams climbed in the dead of night, hammering pitons into the rock, attaching links to them, and fastening ropes to the links. These served as fixed lines to assist those who followed in their ascent of the vertical face of the ridge. The lead climbing teams reached the top around midnight, and battalion units began their ascent in force. By 0400 on 20 February, all three of 1st Battalion's companies and the company from the 2nd Battalion had reached their individual objectives on top of the ridge without being seen. They attacked the defending Germans and completely surprised them. German daylight counter-

> attacks were repulsed and the division's left flank was secured on Riva Ridge opening the way for the rest of the division to accomplish its mission.
>
> The capture of Riva Ridge was the only significant action in which the division put to use its specialized stateside training. Nevertheless, no one would deny that this single mountain operation justified all the demanding training the 10th Mountain Division had undergone.
>
> Adapted from *See Naples and Die*, Robert B. Ellis

GUIDES

4-39. Mountain guides are useful for any operation, particularly *on* terrain above Class 3, which requires accurate judgment and extensive technical knowledge. They act primarily as advisors to unit commanders and assist with planning when technical mountaineering problems affect the tactical scheme of maneuver. They must be experienced in all aspects of mountaineering (usually a Level 3 mountaineer) and capable of ensuring that the force is never beyond the margin of its operational capabilities while operating on mountainous terrain. Mountain guides must possess the skills, knowledge, and experience necessary to develop a true perspective of the obstades and conditions that must be overcome.

4-40. Mountain guides should perform the following functions for the commander:

1. Conduct a terrain analysis of the assigned area.
2. Select the best march routes
3. Identify danger areas, obstades, and hazards.
4. Estimate the effects of elevation on force capabilities.
5. Determine the technical means necessary to employ the force.
6. Estimate the time of unit movements.
7. Develop a movement sketch indicating the azimuth, time, and mountaineering requirements
8. Lead units, usually company-level and above, on difficult terrain.
9. Supervise al l aspects of mountaineering safety.

LEAD CLIMBING TEAMS

4-41. As with all spedal purpose teams, lead dimbing teams have highly skilled soldiers qualified in advanced mountaineering techniques (normally Level 2 mountaineers). To a lesser degree, lead dimbing teams perform many of the same functions as guides. They also accompany a unit over unprepared routes and assist with actually conducting its mission by maintaining and improving mobility. In some instances, lead dimbing teams may operate independently of other forces to accomplish spedfic missions.

4-42. These teams are capable of climbing at night with the aid of night-vision goggies and can conduct operations over any type of terrain. Lead climbing teams should be utilized to install fixed ropes to assist personnel over exposed terrain. They make the most difficult climbs and act as rope leaders. Members of the lead climbing teams must be extremely proficient in the technical aspects of mountaineering, since they select the specific routes to be climbed. This responsibility emphasizes the importance of accurate judgment, since a single mistake could jeopardize the success of the unit's mission. Commanders should consider assigning lead dimbing teams to, or developing an organic capability within:

- Ground reconnaissance elements.
- Forward observer parties.
- Air defense sections.
- Communications sections.
- Security elements.
- Assault elements.
- Sniper sections.

4-43. The number of lead dimbing teams required is dependent on the mission and difficulty of the elevation level or dismounted mobility classification. If more than one point of attack is to be used or a more mobile patrol is needed, it may be necessary to employ additional lead climbing teams. Lead dimbing teams may also assist in the attack of very steep objectives by negotiating the most unlikely avenues of approach. They can be dispatched ahead of the attacking elements to secure the advance at night and during periods of inclement weather. The commander's imagination and their limited availability are the only restraints toward using lead dimbing teams.

4-44. In predpilous terrain, lead dimbing learns can be used alone to conduct small reconnaissance patrols or to form the nudeus of larger patrols. They may also be used to expedite movement of flank security elements over difficult terrain and during poor visibility. At least one team should be attached to each element, depending on the size of flank security and the operational terrain level in which operations take place. When the AO dictates a rate of march on the flanks that is slower than that of the main body, flank security elements should be located adjacent to the advance echelon. Lead dimbing teams are detached as necessary to reconnoiter and hold dominant terrain features on the flanks of the line of march. As the trail element passes and the security position is no longer required, the lead dimbing team's flank outposts joi n the mai n body or proceed forward to another security location.

INSTALLATION TEAMS

4-45. Installation team members are qualified in the construction and maintenance of technical mountaineering systems, referred to as installations, which fadlitate unit movement. These teams deploy throughout the AO. in any weather or visibility conditions, to erect mountaineering installations that overcome obstacles to the movement of friendly forces and supplies.

4-46. In most situations, installation teams should consist of six qualified members, enough to build most installations. Installation teams deploy early and prepare the AO for safe, rapid movement by constructing various types of mountaineering installations (see Figure 4-6). Following construction of an installation, theteam, or part of it, remains on site to monitor the system, assist with the control of forces across it, and make adjustments or repairs during its use. After passage of the unit, the installation team may then disassemble the system and deploy to another area as needed.

- Fixed ropes
- Pulley systems
- Hauling lines
- Rope ladders
- Rope bridges
- Handrails
- Traverse systems

Figure 4-6. Types of Installations

4-47. Although installation teams can assist, commanders remain responsible for establishing a traffic control system and the rate of negotiation to expedite tactical movement and prevent unnecessary massing of personnel and equipment on either side of t he obstacle- Plans for the traffic control system includearrangements for staging and holding areas, as well as movement. A prearranged order of movement, dissemi nated to all elements of the force, facilitates rapid movement. The traffic control plan also includes instruct ions for dispersing forces on the far side once they negotiate the obstacle. The successful negotiation of systems depends on how quickly a force can consolidate on the far side and continue its mission.

4-48. Commanders must carefully consider the proportion of installation team assets allocated to maneuver with those assigned to sustain essential logistics activities. When considering a unit's scheme of maneuver, the number and types of installations depend on METT-TC, emphasizing factors listed in Figure4-7.

- Number of routes required by the maneuver force
- Size of the maneuver force
- Amount and type of maneuver force equipment
- Number of installations required to extend logistics support
- Weather and visibility conditions
- Number of installation teams available

Figure 4-7. Factors Influencing Mountain Installations

4-49. Once operations stabilize, installation teams can direct and assist engineers in the construction of fixed alpine paths. Fixed alpine paths consist of permanent or semi-permanent mountain aids that assist troops in traversing rugged mountain terrain and facilitate the movement of equipment and supplies to and from forward areas.

SECTION II - OFFENSIVE OPERATIONS

4-50. Offensive operations in the mountains vary depending on the degree of restrictions dictated by mountains of different heights and character, but are normally planned and conducted utilizing the movement to contact and the attack. Exploitation and pursuit are conducted, but less frequently than in other environments.

4-51. Mountain operations will most likely be fought to gain control of key or decisive terrain. The goal is to seize objectives that are important for continuing the battle, such as lines of communication, passes, ridges, and choke points. Every attempt must be made to fight from the heights down. Frontal attacks against defended heights have little chance of success and attacks are usually made along the flanks and to the rear of the enemy. Consequently, envelopment becomes the preferred form of maneuver.

4-52. The missions assigned to units operating in a mountainous region remain the same as those in lowland regions. Commanders should place an increased emphasison;

- Limited-visibility operations,
- Mobility and survivability.
- Securing friendly lines of communications while severing those of the enemy.
- Reconnaissance and security,
- Centralized planning and decentralized small-unit actions.
- Environmental factors favoring operations of short duration and violent execution.

PLANNING CONSIDERATIONS

4-53. When conducting offensive operations in the mountains, surprise is often a dominant consideration, rather than a supporting one. Units achieve surprise by achieving superior situational understanding, especially of the terrain, and by using that knowledge to do the unexpected. Friendly forces should conduct bold and imaginative operations to exploit enemy weaknesses or inability to operate in a mountainous environment. A well-trained force can achieve surprise by infiltrating and attacking the enemy's rear or attacking during periods of limited visibility, such as night, rain, or snow. They can use helicopters and their technical mountaineering skills to conduct decisive operations anywhere in the AO. The effects of surprise can be dramatically increased if commanders select objectives in restrictive terrain that decrease the enemy's mobility and ability to react effectively.

4-54, The mountainous terrain increases the threat to concentrated formations. Usually, it is difficult to coordinate all forces by time and location so that they can rapidly support each other and achieve massed effects. The compartmented terrain separates adjacent units, precluding mutual support, and may adversely affect supporting distances. Therefore, it is critical to anticipate the concentration of forces and fires before the battle begins to achieve effective synchronization.

4-55. Commanders usually select decisive points for the attack based, in part, on their ability to seize and control key terrain. Each of these objectives often necessitates the seizure of one or more intermediate objectives. The compartmented terrain and resulting dispersion make it difficult for commanders to visualize the entire AO and complicates command and control, while the terrain often affords the defender excellent observation and decreases the attacker's ability to concentrate forces undetected.

4-56. Once a battle is joined, helicopters are the only rapid means for massing forces on terrain higher than Level I. When weather conditions permit, attack helicopters and tactical air assets are essential because they can move and strike over a large AO in a short time *Therefore,* the suppression of enemy air defense may become a high priority during offensive operations.

4-57. Commanders must personally acquaint themselves with the terrain to the fullest extent possible They combine this knowledge with other factors of METT-TC to develop simple, yet precise plans and orders. As previously discussed, compartmented terrain tends to divide the battle into many isolated engagements that are difficult to control. A complex and inflexible plan wilt not withstand the changing situations associated with this type of decentralized combat. An uncomplicated plan with a well-thought-out intent that is clearly communicated ensures understanding at the lowest level and allows subordinates to exploit battlefield opportunities even if communications fail.

PREPARATION

4-58. The length of the preparatory phase is typically longer in a mountainous environment. An offensive action against an enemy defensive position must be based on thorough reconnaissance and orderly preparation. The primary factor in determining the technique to be used in destroying the enemy position is the strength of the enemy's defenses. The stronger the defenses, the more deliberate the attack must be. In mountainous terrain, evaluation and exploitation of the ground are essential. Commanders must prepare plans of attack that take advantage of the weaknesses found in the enemy's defensive system. In the mountains, a larger number of reconnaissance assets and additional time may be needed to determine the strength of enemy positions on the objective and all surrounding heights, and favorable routes to and past the objective.

4-59. Difficult approach routes should be marked and prepared for safe passage. Easily traversed slopes, broad hills, plateaus, and valley floors, as well as mountainous terrain with well-developed road and transportation nets, permit deployment in breadth. High ranges with ridges and crests leading to the objective require organization in depth with extended lines of communications

4-60. In trackless mountain terrain, company-sized teams usually conduct attacks, tfthearea assigned toa battalion permits, companies should approach the objective separately on multiple routes. In restrictive terrain, adequate maneuver space may not always be available and several units may be required to move along the same avenue of approach. It may even be necessary to conduct shaping operations to seize sections of terrain from which the enemy can dominate the forces' movement and approach.

4-61. The preparatory phase may also indude feints and demonstrations to shape Ihe battlefield. In mountainous lerrain, the defending force has a number of advantages that allow it to defeat an attacking force much larger than its own, such as long-range observation, rugged natural cover, concealment, advance siting of weapons, and operations on familiar terrain. Feints and demonstrations mask friendly operations, expose enemy vulnerabilities, disorganize the defense, and allow the attacking force to engage the enemy discriminately. In a mountain area of operations, the benefits achieved from a successful deception effort may well outweigh the difficulties involved in mounting it.

FORMS OF MANEUVER

4-62 The forms of offensive maneuver are common to all environments, to indude mountainous terrain. While frequently used in combination, each form of maneuver attacks the enemy in a different way, and some pose different challenges to the commander attacking in the mountains.

INFILTRATION

4-63. Infiltration is frequently used in the mountains. The difficult terrain and recurring periods of limited visibility allow for undetected movement. Infiltration in a mountainous environment is used to shape the battlefield by attacking enemy positions from the flank or the rear, securing key terrain in support of the decisive operations, or disrupting enemy sustaining operations. Infiltration is normally conducted using one of threetechniques: movement in small groups along one axis, movement in one group, or movement in small groups along several routes at the same time. Regardless of the technique used, units must move in a covert manner to reduce the chance of enemy contact.

4-64 With movement in small groups along one axis, all members of the force use the best route. Small groups are harder to deled and easier to control, and do not compromise the total force if detected. This technique may require an excessive amount of time and an increased number of guides and lead dimbing teams, and does require an assembly area or linkup point prior to oondud of the dedsive action. If the lead group is detected, other groups may be ambushed.

4-65. The enemy can more easily deted movement in one group. If the force is deteded, the overall mission may be endangered. However, this technique has no reassembly problems, since reassembly is not required. Everyone uses the same route, easing navigation and redudng the number of guides and lead dimbing teams required. A large force can fight out of a dangerous situation more easily than a small one This technique minimizes coordination problems with other infiltrating units.

4-66. Movement in small groups along several routes at the same time has several advantages. It avoids putting the total force in danger and is less likely to be seen. Il forces the enemy to react in many locations and makes il harder for him to determine the size of the force or its mission. Groups travelling over severely restrictive terrain may have significant assembly, control, and sustainment difficulties, and may require more guides, lead climbing teams, and installation teams.

ENVELOPMENT AND TURNING MOVEMENT

4-67. The envelopment and its variant, the turning movement, are used extensively in mountain operations. Both forms of maneuver seek to avoid the enemy's strength and attack the enemy at a decisive point or points where the enemy is weakest or unprepared, and both rely on superior agility. In the mountains, the ability to react faster than the enemy may be achieved through greater mountaineering skills, using airborne and air assaults, and, depending on the specific AO, amphibious assaults.

PENETRATION

4-68. Mountainous terrain normally makes penetration extremely dangerous or impossible because of the difficulty in concentrating overwhelming combat power in the area of penetration. Due to mobility restrictions, it is also difficult to develop and maintain the momentum necessary to move quickly through a gap and on toward the objective. The area of penet ration is always vulnerable to flank attack, and this vulnerability significantly increases in mountainous terrain A penetration may be useful when attacking an enemy that is widely dispersed or overextended in his defense. If a penetration must be conducted, flank defensive positions must be eliminated before the initial breach of enemy positions. Successful penetration of a defensive positron in restrictive mountainous terrain requires using limited visibility, stealth, and covered and concealed terrain at selected breach points.

FRONTAL ATTACK

4-69. Frontal attacks in hilly or mountainous areas, even when supported by heavy direct and indirect fires, have a limited chance of success. Mountain terrain adds to the relative combat power of the defender. The frontal attack exposes the attacker to the concentrated fire of the defender while simultaneously limiting the effectiveness of the attacker's own fires. In the mountains, the frontal attack is an appropriate form of maneuver to conduct as a shaping operation designed to fix a force, while the decisive operation uses another form, such asthe envelopment, to defeat the enemy.

MOVEMENT TO CONTACT

4-70. The fundamentals regarding movement to contact also apply in a mountainous environment with the added likelihood of surprise attack and ambush. Limited mobility and dependence on restrictive terrain make it difficult to rapidly deploy from the movement formation. Plans and movement formations and techniques should be based on maintaining fiexibility and providing continuous security.

4-71. During a movement to contact, the advance guard normally advances in column, moving continuously or by bounds, until it makes contact. While requiring less physical exertion, movement along the topographical crest of a ridgeline increases the possibility of enemy observation and should normally be avoided. Given adequate concealment, this exposure may be reduced by moving along the military crest. Ridgelines and crests can often provide a tactical advantage to the force that controls them. Their control may allow rapid movement from one terrain compartment to another and afford excellent observation into lower terrain levels. In all cases, commanders must address the control or clearance of ridgelines that dominate their planned avenues of approach.

4-72. The main body should never be committed to canalizing terrain before forward elements have advanced far enough to ensure that the main body will not become endrded. This is a critical fador when employing mixed heavy and light forces that have sharp differences in operational tempo. Combat service support must be decentralized and readily available to sustain the combat elements. Major terrain compartments may physically separate maneuver units moving as part of a larger force. Continuous reconnaissance to the front and flank security is essential to prevent the enemy from infiltrating the gaps between units.

4-73. As the enemy situation becomes better known, commanders may shorten the distance between elements to decrease reaction time, or they may begin to deploy in preparation for the attack. Lateral movement between adjacent columns is frequently difficult or impossible. However, every attempt should be made to maintain at least visual contad. Commanders must emphasize the use of checkpoint reporting, contact patrols, and phased operations to coordinate and control the movement of the overall force. Control measures should not be so numerous as to impede operations and stifle initiative. Proper control ensures that units and fires are mutually supporting, objectives arecorrectly identified, and unitsare in position to attack.

ATTACK

4-74. Speed, flexibility, and surprise, normally advantages enjoyed by the attacker, are limited by restridive terrain and the defender's increased ability to see and acquire targets at greater distances. These limitations make it difficult for units above the company team level to conduct a hasty attack against prepared positions. In the mountains, commanders usually need more time to coordinate fire support, pick routes to prevent enemy observation and detedion. and select control measures to coordinate and control the operation. Therefore, deliberate attacks requiring a detailed scheme of maneuver and well-developed fire support plan become the norm at battalion-level and above. Since daylight contributes to the defender's ability to see and, thereby, reduces the attacker's chances of success, commanders should seek opportunities to exploit the advantages of limited visibility. Although these conditions slow movement even more and make coordinating forces more difficult, they decrease the enemy's ability to accurately sense what is happening and react effectively.

4-75. In planning and conduding the attack, commanders should recognize that the enemy will generally seek to control the valleys and trail networks, induding adjacent slopes and high ground. Defenses normally be anchored around obstades, and long-range, di rect fire weapons employed in poorly t rafficable terrain, often on slopes and protruding high ground. The enemy will attempt to engage the attacker in the valleys and low ground with flanking fires and artillery, often in a direct fire mode. Commanders must analyze the terrain to determine not only how the enemy will organize his defensive positions, but also how the terrain might contribute to the enemy's ability to counterattack As friendly forces attempt to deploy for the attack, the enemy, using his advance knowledge of the terrain and prepared routes, may maneuver forces to counterattack from the flank or rear.

4-76, All terrain features that can be occupied by even a small enemy force should be secured. I n many instances, overwatch positions may not be readily available within the range capability of organic weapons. Infiltration, technical dimbing, and extensive breaching may be required to position weapons to support the assault. On many occasions artillery support, espedally in high mountains, may not be available. In other instances, commanders may need to identify intermediate objectives for maneuver forces based on the need to ensure that artillery units have suitable, secure firing positions to range the enemy and support the attack. As in all environments, commanders must identify fire support requirements and allocate fires based on the ability to support and available ammunition. Because resupply may be limited and extremely difficult, they may need to place restrictions on the amount of ammunition expended on spedfictargets

4-77. Fire and movement during the assault are extremely difficult. In situations where machine guns can be positioned effectively, a rifle platoon can provide itself with support from a flank or from a height. However, during an assault up a slope, supporting fire cannot come from an overwatch position and must originate from the flanks or through gaps between the assaulting soldiers. Control is difficult to maintain when the assault is in steeply rising terrain. Commanders must pay speda! attention to the dangers of fratricide.

4-78. Fire and movement are easier in an assault over a downward slope Down-slope assaults often have the advantage of good observation, but dead spaces and intervening terrain may reduce the effectiveness of supporting fires. Defensive positions laid out by a skillful enemy on a reverse slope significantly increase the effect of unfavorable down-slope conditions. This type of defense compels the attacking force to position its supporting weapons and observation posts on exposed crests. In this sit u at ion, support dements must be positioned to avoid terrain masking and crest dearancs problems.

4-79. Breaching obstades and preparing bypass routes that allow the assault force to move into the defensive position must bean integral part of the commander's plan. In rugged terrain, man-made obstades that are covered by fire create a particularly dangerous and formidable barrier. Command and control of a covert, in-stride, deliberate, or assault breaches is more difficult than in open terrain, and mobility support is extensive if the obstade cannot be reduced. Assaults in mountainous terrain almost always involve preparing routes that allow the assault force to rapidly move over difficult natural obstades and into the objectives.

4-80. Commanders should maintain a strong reserve, if possible. In the mountains, as elsewhere, commanders can use their reserves to restore the momentum of a stalled attack, defeat enemy counterattacks, and exploit success. Reserves must becarefulty positioned and organized so difficult terrain, limited road networks, or unpredictable weather does not delay their arrival.

Once committed, commanders make every effort to reconstitute another reserve from available units.

4-81. An attack should not be halted on a summit or on a ridgeline objective, which enemy artillery and mortar fire will likely target. Reorganization is generally best conducted well forward of a crest line on the next suitable slope. Commanders must ensure that the enemy is not allowed the opportunity to counterattack to recapture key terrain. Rapid adjustment of positions and coordination with flanking units are essential. Support weapons, especially mortars, should be brought forward as quickly as possible. Helicopters are useful for this purpose and may be used for backhaul of casualties

COUNTERATTACK

4-82. Counterattacks in the mountains must exploit the aspects of terrain that impair enemy momentum and make it difficult for him to mass and maneuver. Obstructing terrain that canalizes movement and restricts mobility significantly increases the potential for counterattacks. In planning a counterattack, the commander must carefully consider the enemy's weaknesses or inability to operate in a mountainous environmen l. A counterattack, even on a very small scale, can have a decisive impact in mountainous terrain.

RAID AND AMBUSH

4-83. The restrictive terrain also affords increased opportunities to conduct raids and ambushes. These operations should take advantage of limited visibility and terrain that the enemy may consider impassable. In steep terrain, movement time increases significantly, and only light equipment can be taken. The force should use special climbing techniques to negotiate the difficult routes during limited visibility. Commanders must carefully consider the routes and methods used for extraction to ensure that the combat force does not become isolated after executing the mission. They can ensure a successful operation by avoiding detection through proper movement techniques and by skillfully using natural cover and concealment. It may be necessary to reposition some indirect fire support assets to cover dead space or use attack helicopters and close air support. The ambush or raid commander must know in advance if supporting fires can not cover his routes to and from the objective.

DEMONSTRATIONS AND FEINTS

4-84 Because maneuver space is usually limited or confined and restricts the number of avenues of approach for heavier forces, deception plays an important part in the mountain battle To mislead the enemy regarding friendly intentions, capabilities, and objectives, commanders should plan systematic measures of deception.

EXPLOITATION AND PURSUIT

4-85. In a mountainous environment, exploitation and pursuit operations must be conducted discriminately and the mountain commander must always prepare for success. A battalion may exploit its own success to a limited extent, but it normally participates in the exploitation as part of a larger force. Air assault and attack helicopter units can be used to augment exploitation and pursuit operations. The exploiting commander must compensate for the ground mobility restrictions imposed by terrain and weather. Speed can best be achieved by isolating enemy positions with the smallest force possible Engineer support should be well forward with the necessary equipment to allow combat troops to maintain momentum and avoid delay by enemy obstacles. The commander must be careful to prevent overextending either the exploiting force or its sustaining logistics. A withdrawing force is capable of establishing numerous strong points and firing positions on heights that allow it to qu i ckl y di ssi pate the com bat power of t he expl oi t i ng f or ce.

MOTTI TACTICS

4-86. Motti tactics are presented here to demonstrate how forces can exploit superior mobility skills and knowledge of the mountainous terrain and environment to defeat the enemy. The Finns developed these tactics during the Finnish-Russian War in 1939-1940. They are characterized by attacks on rear areas, bivouac sites, and command posts.

4-87. The Finnish word "motti" means a pile of logs ready to be sawed into l umbo; in effect, setting the conditions so that a larger force can be defeated in detail. These tactics were most successful in the forested areas of Finland during the arctic winters. During the Finnish-Russian War, the Soviets were neither prepared for, nor trained for, warfare under such conditions. They were almost totally trail-bound, with few ski troops. In the 1980s, the Soviet Union experienced similar difficulties in the mountains of Afghanistan, In both instances, the road and trail-bound nature of their forces and their basic tactics left them vulnerable to motti tactics in mountainous terrain.

4-88. Generally, the force utilizing motti tactics never becomes decisively engaged. It disrupts the enemy's supply lines, denies him warmth and shelter, infiltrates his bivouacs, and destroys his rear areas to the point where he must remain in a high state of alert. These attacks, in combination with the environment, help to destroy the enemy's will to fight Commanders should not only develop a thorough understanding of how to apply these tactics, but also understand the conditions that may leave their own forces vulnerable to its use (see Figure 4-B on page 4-24).

Motti tactics generally follow the sequence of:

1. Locating and fixing the enemy.
2. I solating the enemy.
3. Attacking to defeat or destroy the enemy

4-89 Reconnaissance is conducted to locate an enemy force moving in or toward an area that will restrict his movements to roads, trails, or linear terrain. Once identified, the force must be fixed so that it presents a linear target along the axis of advance to which it is bound. This is accomplished using obstacles and a series of squad and platoon sized ambushes and raids Obstacles may be natural, such as snow, crevices, deep mud, steep terrain, and water obstacles, or man-made, for example mines, landslides, avalanches, or destroyed bridges.

4-90. The ambushes and raids not only fix the enemy, they also disturb his composure, create an air of uncertainty, and prevent uninterrupted sleep and rest. Friendly units attack the enemy from the high ground. They make maximum use of night vision deviceSr as well as the difficult restrictive terrain. They avoid enemy security and interdict his operations. As a further result of these actions, the enemy is compelled to use more forces on security tasks. Unless the enemy can be easily defeated or destroyed, the attacking force rapidly withdraws after forcing the enemy to deploy. In general, this series of attacks confuses the enemy as to the friendly unit's exact location and intent, and slows his decision-making cyde so that he reads ineffectively to subsequent operations.

FORCES CAN USE MOTTI TACTICS WHEN THEY:

- Have superior technical mobility skills necessary to negotiate Class 4 and 5 terrain
- Are able to operate effectively in a noncontiguous area of operations with limited support, and despite temperature extremes and inclement mountain weather
- Are able to navigate in high mountainous terrain, dense vegetation, darkness, storms, and fog while making good use of available cover and concealment
- Maintain the element of surprise

FORCES ARE VULNERABLE TO MOTTI TACTICS WHEN THEY:

- Operate within noncontiguous areas of operations
- Have limited mobility skills restricting their movements to roads, trails, and Class 1 and 2 terrain
- Have inadequate reconnaissance and security

Figure 4-8. Conditions Affecting the Use of Motti Tactics

4-91. The attacking force then isolates the enemy into smaller groups. Once isolated, the friendly force maneuvers to envelop and attacks to defeat or destroy the isolated elements. As the enemy exhausts himself in an effort to break out, the attacking force may regroup and repeat the sequence. It is imperative that the attacking force seal off the enemy and keep avenues of approach closed, and not ignore the threat to its flanks, which may increase as the attack progresses.

4-92. Overall, motti tactics wear the enemy down to a point where he is vulnerable to more direct attacks or to the point where it is no longer beneficial or feasible to continue operations in the area. Motti tactics employed alone only prove decisive over a long period of time, depending on the enemy's capabilities, strength, and resolve. Based on METT-TC. friendly forces normally must increase the operation's tempo to gain a quick, decisive outcome. Sti 11.

these type tactics may complement other more direct offensive operations in support of the overall pfan.

SECTION III - DEFENSIVE OPERATIONS

4-93. In the mountains, more so than in the lowlands, the strength of the defense depends on its selection and use of key and decisive terrain. Key and decisive terrain provides the defender-and usually denies the attacker-excellent observation and fighting positions. Reinforcing obstades significantly enhance the natural obstacles of rugged mountainous terrain.

4-94. The immediate objective of a mountainous defense is to deny the enemy access to key terrain that helps him conduct further operations Therefore, it is necessary to defend in terrain that restricts and contains the enemy, as well as control the high ground that dominates this terrain. The effects of rapidiy changing weather, visibility, and mountain hazards must be continually assessed. The terrain provides the defender with cover, concealment, and camouflage that can deceive the enemy regarding the strength and dispositions of friendly forces. The advantages of knowing the terrain, having fortified positions, siting weapons in advance, stockpiling supplies, and identifying or preparing lateral trail networks favor the defense. They allow the defender to shift forces on the ground more rapidly than the attacker. Delaying operations are particularly effective in the mountains and can be accomplished by a smaller force. These advantages combine to make the mountai ns an ideal place for defensive operations. Regardless of the scale of defensive operations, key factors in achievi ng success in the mount ai ns are having good observation and aggressive reconnaissance, white denying the same to the enemy.

PLANNtNG CONSiDERATIONS

4-95. Defending commanders must develop flexible pEans for control of fire, maneuver, communications, and logistics. Initially, the attacker has the initiative and decides where and when combat will take place. The defender must be agile enough to maintain control of the heights, strike effectively, and shift his effort quickly without losing momentum and flexibility. Tactical flexibility depends on planning in detail, organizing in depth, and retaining an adequate, mobile reserve

4-96. Although the mountains generally allow observation at greater dislances, intervening terrain features and weather often prevenl commanders from seeing the area of operations beyond the area to their immediate front and flanks. Consequently, commanders normally allocate more assets for reconnaissance and security, echeloned in depth and in height, to ensure that they are able to sense all aspects of the AO and gain the time needed to decisively apply combal power.

4-97. Commanders must prevent the enemy from concentrating overwhelming combat power against isolated sections of their defense. The restrictive terrain is one of the primary advantages of the mountain defender, as it interferes with the attacker's synchronization, canalizes his movement, and impedes his ability to maneuver. However, unless oommanders carefully analyze the terrain from both the friendly and enemy viewpoints, to indude the horizontal and vertical perspectives, they leave themselves vulnerable to infiltration and possible attack from the flanks and rear along difficult and unexpected routes.

4-98. In the mountains, commanders usually organize for a perimeter defensetobe prepared to defeat the enemy from any approach, to indude those that may appear impassable. Although preparing for an all-around defense, they should weight a portion of the perimeter to cover the most probable direction of enemy attack. Rocky terrain may make it more difficult to prepare defensive positions and rapidly changing weather may halt preparations altogether. If suffident forces are not available, the commander must economize in some areas and rely moreheavily on prepared positions, to ind ude alternate and supplementary positions, obstades, and well-planned indired fires to cover gaps and dead space.

4-99. The width of an area to be defended depends mainly on the degree to which terrain is an obstade. Terrain that significantly restricts enemy movement tends to favor a larger AO. Normally, an area should be approximately as deep as it is wide, and may indude the entire length and surrounding heights of a valley. Ridges that run at right angles to the enemy's direction of attack also permit increased width with less depth. A defense in depth is required in valleys that run in the direction of the enemy's attack. In either case, it is essential to have forces deploy on the dominating heights that control approach routes.

4-100. Ideally, reserves should be mobile enough to read to enemy adion in any portion of the perimeter. Less mobile reserves are positioned to block the most dangerous avenues of approach and assigned on-order positions on other critical avenues. Sharply compartmented terrain may require the creation of more than one reserve. Helicopters may be used to deploy reserves, but their use depends on the availability of suitable, secure LZs and favorable weather conditions.

PREPARATION

4-101. The process of preparing the defense must begin with a thorough reconnaissance. Preparations for a mountain defense require more time than in other terrain, and as units arrive they must begin immediate preparation of their defensive positions. In some instances, technical mountaineering skills may be needed to establish effective security and to emplace crew-served weapons properly. However, commanders must weigh the advantages gained from these inaccessible positions against difficulties in repositioning and resupply. Preparations for the defense must alsoindudeinstalling communications, stocking forward supply points with particular attention to Class IV. empladng medical elements, adjusting air defense coverage, and arranging for security of installations in the rear area. Commanders must ensure that time is available to develop alternate routes and positions, rehearse and time movements between positions and along routes, and rehearse counterattacks.

4-102. Commanders must seek every opportunity to recapture the initiative from the attacker and transition to offensive operations. Preparations for a counterattack in the mountains must include caching ammunition, preparing counterattack positions and routes to attack downhi 11, identifying crew-served weapon positions, and establishing rally pointsthat are usually on the reverse slope.

ORGANIZATION OF THE DEFENSE

4-103. Defensive operations in the mountains derive their strength, balance, and freedom of action from the effective use of terrain. The mobility restrictionsfound in mountainous areas, combined with the necessity to hold dominating ground, dictates that an area defense be used. Mountain defenses use security forces, continuous reconnaissance and combat patrols, as well as numerous observation posts. The mountain AO is usually organized into a security area, main battle area (MBA), and rear area.

SECURITY OPERATIONS

4-104. While a screening force is often thought to be the most preferable form of security in extremely rugged mountainous terrain, all forms of security operations, to include guard, cover, and area, may be employed effectively in a mountain AO based on the factors of METT-TC with particular emphasis on:

- Forces available for security operations.
- Abilitytomaintainamobility advantage.
- Size of the security area and the number of avenues of approach.
- Likelihood of enemy action,
- Size of theexpected enemy force.
- Amount of earlywarningand reaction time needed.

4-105 A screen primarily provides early warning to the protected force and is usually an econorny-of-force measure. The compartmented nature of mountainous terrain often serves to create multiple gaps and exposed flanks The rugged terrain may also serve to restrict the movement of not only advancing enemy forces, but also the movement and mobility of larger friendly security forces. In these instances, commanders may choose to use minimum combat power to observe, identify, and report enemy actions at these locations, and engage and destroy enemy reconnaissance within the screening force's capability. The screening force may be able to avoid decisive contact by withdrawing into restrictive terrain that forces the enemy to utilize difficult climbing techniques if he continues the pursuit.

4-106. In the mountains as elsewhere, the screening force should adjust to the enemy advance and continue to screen as far forward as possible, even though elements of the force may have to withdraw. Retention of selected forward positions may allow surveillance and targeting forward of the MBA, upsetting the enemy's coordination. By allowing the enemy to bypass advance positions, the screening force can facilitate counterattack to the front of the forward edge of the battle area (FEBA) by providing observation of. and access to. the flanks and rear of attacking forces.

4-107. If a significant enemy force is expected or a significant amount of time and space is required to provide the required degree of protection, commanders usually resource a guard or cover mission instead of a screen. As long as trie security force can maintain a mobility advantage over the enemy, it can effectively delay and attack the enemy force by using obstacles and the restrictive terrain to its advantage. Although utilizing a greater proportion of his combat power, the appropriate use of a guard or cover force should provide the mountain commander greater depth in his security area and the ability to defeat, repel, or fix lead elements of an enemy ground force before they can engage the main body with direct and indirect fires.

4-108. No matter the type of security used, defending forces must prevent enemy infiltration by carefully positioning observation posts (OPs) and conducting continuous patrols and ambushes. Combat reconnaissance patrols and other intelligence gathering assets observe the enemy as far ahead of friendly positions as possible and report his strength and composition, as well as his route of movement. To accomplish this, reconnaissance patrols may need to rely heavily on technical climbing skills. Ground surveillance radar and unattended ground sensors can be used effectively, but the defender must besureto cover all gaps and dead spaces. The defender must make best use of his time to study the ground and determine ail possible infiltration routes.

MAIN BATTLE AREA

4-109. In rugged mountainous terrain, it may be difficult to maintain mutual support and overlapping observation. Elements should be employed to man observation posts, assist the passage of security forces into the MBA. cover obstacles and avenues of approach by fire, screen gaps between defensive positions, and ambush enemy infiltrators.

4-110. Defensive positions along ridges or domi nating heights should include as much of the forward and reverse slopes as possible to add depth and all-around security. The actual size of unit positions is terrain-dependent. At a minimum, fighting positions and observation posts should be echeloned vertically, as well as in depth. When defending a mountain valley, forces should establish fighting positions that are located on adjacent heights and in depth to permit coveringthevalleywith interlocking fire. These positions must also be anchored to restrictive terrain or adjacent defensive forces to prevent enemy envelopment. In wooded terrain, defensive positions may be organized on theforwardedgeofthe woods, as well as on commanding heights. Obstacles should be widely employed to slow or stop enemy movement throughout.

4-111, Mountain warfare demands that forces conduct an aggressive defense. Defending units must infiltrate enemy units and attack headquarters, supply lines, and rear areas. Smaller patrols and OPs should be deployed well forward to direct artillery fire and attack aircraft on targets of opportunity, and to conduct personnel and antiarmor ambushes. Disruption operations should be conducted to force the enemy to deploy additional assets to protect lines of communication and delay and upset preparations for the attack. In the mountains, enemy forces can frequently be isolated if they are discovered in time and reserves are effectively placed and highly mobile.

4-112. In the defense, the reserves' primary purpose must preserve the commander's flexibility In mountainous terrain, the commander may need to rely on the reserves as his principal means of restoring his defense's integrity or exploiting opportunities through offensive action. Because of the difficulties of movement, small reserves may be located near primary defensive positions, ready for immediate counterattack. This type of small, responsive counterattack may be much more effective than a large-scale, major counterattack. It can catch the enemy exhausted after an uphill assault and before position consolidation. Large, centrally placed counterattack forces are normally unableto intervene in time unless the terrain permits mounted movement, or sufficient helicopter lift assets are committed to the reserve force or made rapidly available.

REAR AREA

4-113. To minimize the vulnerability of sustaining operations and extended lines of communication, command and control, as well as support operations, in the rear area must be dispersed, redundant, and as far from potential enemy approaches as possible. Because of limited space available in the rear area, the commander must be careful in selecting and locating positions for combat service support activities. These positions are likely to be confined to small valleys. Therefore, they are high-priority targets for enemy artillery and air attack or raids by small combat patrols, particularly at night or in bad weather. When possible, combat service support elements must avoid the most obvious positions and occupy atypical sites. However, they should be in the vicinity of a defined road network and an air loading area, even if the network or area is within Level 11 Locating base defenses at Level II elevations may allow more access to supply bases for air resupply during inclement weather, such as the heavy fog often encountered in valleys and at lower elevations

4-114. A perimeter defense is planned for each combat service support unit within the defensive area. Defensive positions should be selected on the dominating high ground. Sensors, OPs. and radars are used to cover avenues of approach and gaps between positions. Rear area forces must routinely conduct patrols and ambushes around the perimeter, especially at night and during other periods of limited visibility. Air defense assets should be located to protect rear area facilities. Tactical combat forces (TCP) must be prepared to respond rapidly to rear area threats and should be prepared to move to any of their objectives by multiple routes. However, units within the rear area must not fall into the trap of relying solely on a TCP for their security. No matter how well-organized or mobile the TCP, rear area units must provide their own well thought-out and active security measures, even at the cost of a reduced ability tosustain the force.

REVERSE SLOPE DEFENSE

4-115. Reverse slope defenses apply particularly well to mountain operations and pursue offensive opportunities through surprise and deceptive operations by defending in a manner for which the enemy is unprepared. This defense seeks to reduce the effects of massed indirect fire from mortar, artillery, and dose air support, and draws the battle into the small arms range of infantry weapons. The overall goal is to make the enemy commit his forces against the forward slope of the defense, causing his forces to attack in an uncoordinated fashion across the exposed topographical crest. Once this type of defense is employed, subsequent use may be of limited value, due to the loss of the key element of surprise.

4-116. All or partsof thedefending force may use reverse slope techniques. In many instances, mountainous terrain favors a defense that employs combined forward and reverse slope positions to permit fires on enemy approaches around and over the crest and on the forward slope of adjacent terrain features. Key factors to this type of defense are mutually supporting covered and concealed positions, numerous natural and man-made obstacles, theability to bring fire from all availableweapons onto the crest, and a strong and mobile counterattack force.

4-117. The reverse slope defense is organized so that the main defensive positions are masked from enemy observation and direct fire by the topographical crest (see Figure 4-9). It extends rearward from the crest only to the maximum effective range of small arms fire. Observation and fires are maintained over the entire forward slope as long as possible to continue to destroy advancing enemy forces and prevent him from effectively massing for a final assault. A successful reverse slope defense is based on denying the topographical crest to the enemy, either by fire or by physical occupation. Although the crest may not be occupied in strength, control of the crest by fire is essential for success. For more detailed discussions of the reverse slope defense, see FM 3-100.40 and FM 3-21.30,

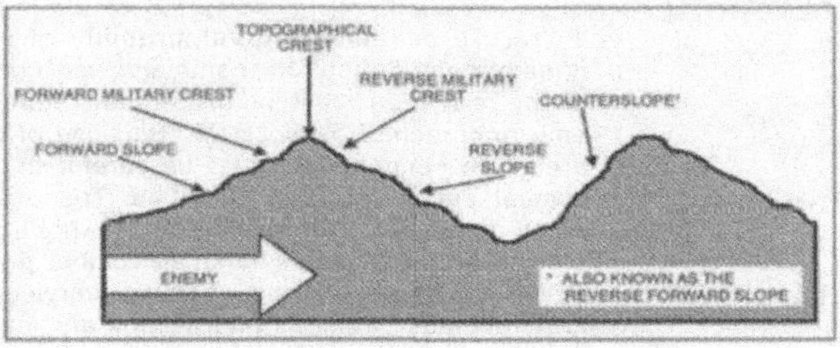

Figure 4-9. Reverse Slope Defense

RETROGRADE OPERATIONS

4-118. Retrograde operations of delay and withdrawal can be conducted in mountainous terrain with fewer assets because of the mobility difficulties of an advancing enemy. Delaying operations are particularly effective in the mountains. Numerous positions may exist where elements as small as a machine-gun or sniper team can significantly delay a large force. When conducting retrograde operations in mountainous terrain, the friendly force must accomplish several tasks.

- it must make maximum use of existing obstacles- However, the addition of relatively few reinforcing obstacles, such as the antitank mining of a route with very steep sides, often increases the value of existing obstacles.
- The force must conduct detailed reconnaissance of routes to rearward positions. Routes of withdrawal are not as numerous in mountainous terrain and often do not intersect as they do on flat terrain. These

factors complicate subsequent link-up operations and necessitate meticulous planning.

- It musl protect the flanks and rear to prevent encirclement, particularly by air assault. There are only a few LZs and they can significantly influence the outcome of a battle. At a minimum, they must be covered by fire.

STAY-BEHIND OPERATIONS

4-119. The compartmented terrain in the mountains lends itself tothe employment of stay-behind forces as a tool for offensive action. Stay-behind operations involve the positioning of friendly elements within operational areas before the enemy advances through the area. Stay-behind forces conceal their location and allow themselves to be bypassed as the enemy advances. (Figure 4-10 outlines the important tasks that stay-behind forces can accomplish for the mountain commander.)

- Attack the enemy throughout the depth of his formations
- Disrupt the cohesion of the enemy offense by interrupting lines of communication and logistics
- Detract from the enemy's main effort by forcing him to allocate combat forces to rear areas
- Provide immediate intelligence
- Call for and control indirect fire and close air support

Figure 4-10. Tasks for Mountain Stay-Behind Forces

4-120. Stay-behind forces may be positioned forward of the FEBA, in the MBA, and not participate in the initial fight, or, under certain conditions, in the MBA after the fighting has started. When planning for stay-behind operations in the mountains, commanders must consider the following:

- Stay-behi nd forces should be a combined arms force that indudesengineers.
- Indirect fire support or dose air support should be available throughout their operations.
- Return routes must be well planned and reconnoitered in advance. Exfiltration, regardless of element size, should follow covered and concealed routes, and rally points should be designated forward of and behind the lines of friendly forces. Reentry must be carefully coordinated to prevent fratridde.
- After an attack, stay-behind forces may be unable to reach a hide position to subsequently return to friendly lines by exfiltration. If this occurs, they must be prepared to condud a breakout,
- The tactical situation and logistics supplies that were stockpiled or cached in the AO have an impad on the length of time stay-behind forces remain in enemy territory.

Chapter 5

Logistics and Combat Service Support

Mountainous terrain poses great challenges to combat service support (CSS) forces and complicates sustaining operations. Existing roads and trails are normally few and primitive, and cross-country movement is particularly demanding. Highways usually run along features that have steep slopes on either side, making them vulnerable to disruption and attack. Rivers become major obstades because of rapid currents, broken banks, rocky bottoms, and the lack of bridges. Landslides and avalanches, natural as we! l as man-made, may also pose serious obstacles to CSS operations. Mountainous areas have wide variations in climate and are subject to frequent and sudden changes of weather that may preclude reliance on continuous aviation support. Together, these conditions compound the obstacle-produci ng effects of mountainous terrain and create major challenges for the CSS planner. Therefore, the forward distribution of supplies may depend upon the knowledge, skill, and proficiency of CSS personnel in both basic mountaineering and aerial resupply operations.

CONTENTS

HISTORICAL PERSPECTIVE

The Importance of Lines of Communications: The Satukandav Pass (Soviet-Afghan War, November-December 1987)

One of the characteristics of the Soviet-Afghan War (December 1979 - February 1989) was the attempt by both sides to control the other's lines of communications (LOCs). In an effort to deprive the guerrillas of their source of sustainment, the Soviets used various methods to drive the rural population into exile or into cities. For their part, the Mujahideen regularly interdicted supply routes through the establishment of blocking positions and vehicular ambushes. In some regions, they were able to effectively interdict supply routes for weeks, months, and even years at a time. The Soviet main supply route was a double-lane highway network winding through the rugged and inhospitable Hindu Kush Mountains. The continued Soviet presence in Afghanistan depended, in large part, on their ability to keep the roads open. Therefore, much of heavy Soviet combat was a fight for control of this road network, with this control often changing hands during the course of the war.

In the fall of 1987, the Mujahideen had established a series of blocking positions that severely limited the supply of weapons, ammunition, and food to Soviet forces in the Khost district. In response, the Soviets planned and

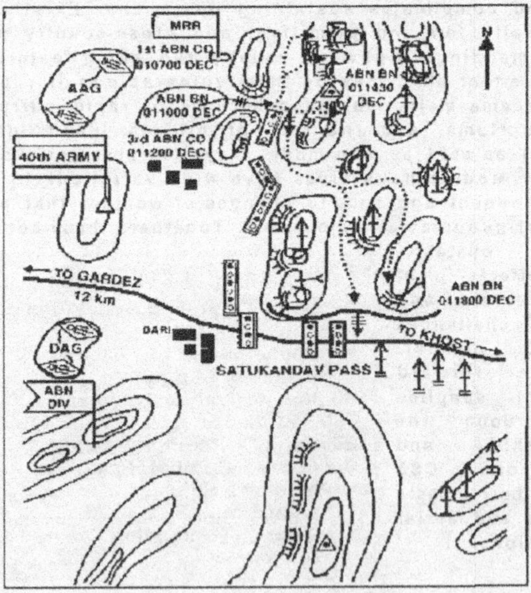

Figure 5-1. Satukandav Pass

conducted Operation Magistral, "main highway," to open the LOCs (see Figure 5-1). The guerrilla forces had established strong positions in the Satukandav Pass, virtually the only way through the mountains between Gardez and Khost. For the operation, the Soviets massed a motorized rifle division, airborne division, separate motorized rifle regiment (MRR), separate airborne regiment, various 40th Army regiments, special forces, and other subordinate units, and regiments from the Afghan Armed Forces. On 28 November, in order to determine the location of Mujahideen positions, particularly air defense systems, the Soviets conducted a ruse in the form of an airborne assault using dummy paratroopers. When the Mujahideen fired at the dummies, Soviet artillery reconnaissance was able to pinpoint enemy strong points and firing positions. The Soviets hit these positions with air strikes and a four-hour artillery barrage. The next day, however, an MRR failed to make its way up the foothills to seize the dominant terrain along the crest, and suffered heavy casualties. The 40th Army Commander, General Gromov, nevertheless, decided to continue to press his advance using the 1st Airborne Battalion and a battalion of Afghan commandos. On 1 December, two airborne companies captured key terrain and used this to support the decisive operation against the dominant peak to the south. This flanking attack took the Mujahideen by surprise and they began to withdraw. While calling in artillery fire on the retreating guerrilla forces, primarily on the reverse slope and along the probable avenue of approach for the commitment of enemy reserves, the Soviet battalion commander used this hard-won, key terrain to support a simultaneous, two-prong attack to the south toward the Satukandav Pass. Now it was the Soviets who were in a position to cut off supplies, especially fresh drinking water, from the Mujahideen. The latter were forced to withdraw, and the two battalions captured the pass. However, while the operation itself was a success, Soviet and Afghan Army forces could keep the road open for only 12 days, after which the Mujahideen once again cut off the supply route to Khost.

Both sides recognized the vital importance of LOCs, and this shift of LOC control was a constant feature throughout the entire duration of the Soviet-Afghan War. The Mujahideen's ability to interdict the LOCs prevented the Soviets from maintaining a larger occupation force there, a key factor in the eventual Soviet defeat.

Compiled from *The Other Side of the Mountain* and *The Bear Went Over the Mountain*.

SECTION I - PLANNING CONSIDERATIONS

5-1. Mountainous areas of operations dictate that commanders foresee needs before demands are placed upon CSS personnel The main logistical differences between mountain operations and operations in other terrain are a result of the problems of transporting and securing material along difficult and extended lines of support Logistics support must emphasize a continuous flow of supplies to specific locations, rather than the build-up of stocks at supply points along the main supply routes. Supply point operations alone are insufficient; the proposed support structure must plan for redundancy in the ability to distribute supplies directly to units operating from predetermined supply routes. If possible, commanders should plan to use multiple supply routes designed to support maneuver elements moving on separate axis.

5-2. Commanders must be concerned not only with the sustainment of current operations but also with the support of future operations. A detailed logistics preparation ol the theater (LPT) to identify the potential lines of communication plays a major part in determining the conduct of CSS operations. A detailed reconnaissance should be conducted to determine:

- The type and maximum number of vehicles that the road network can support in the area. New roads may need to be constructed or improvements made to existing ones to support protracted operations in isolated areas.
- Classification of bridges.
- Suitable sites for drop zones (DZs), loading zones (LZs), and short, tactical airstrips.
- Availability of water sources.
- Availability of local resources, facilities, and service and support activities.

5-3. Because of terrain constraints, it may be necessary to disperse support units over a wider area and ensure that supplies are positioned closer to supported units. Dispersion reduces vulnerability of CSS assets, which also creates problems with command, control, and security CSS units are often high-priority targets, and must ensure adequate protection against ground and air attacks.

5-4, In mountainous terrain, battalion CSS elements are normally echeloned into combat and field trains to increase responsiveness, provide adequate space, and decrease the logistics footprint. Combat trains are routinely located in ravines or valleys on the rear slope of the terrain occupied by the unit. This permits the personnel officer (S1) and logistics officer (S4) to operate in close proximity to the tactical operations center (TOC), and allows them to keep abreast of unit requirements,

5-5. In the mountains, unresolved logistical problems can quickly lead to mission failure. Ground operations may increase fuel consumption rates of individual vehicles by 30 to 40 percent, requiring more frequent resupply operations. The operation of equipment in mountainous terrain has proven that maintenance failures far exceed losses due to combat, and most breakdowns can be attributed to operator training. Air operations are characterized by a significant increase in lift requirements; however, increased elevations decrease aircraft lift capabilities,

5-6. Commanders must carefully consider combat loads in the mountains, based upon a thorough mission analysis. Excess equipment and supplies reduce the efficiency of the individual soldier and seriously impede operations. In steep terrain above 1.500 meters (5,000 feet), soldier loads may need to be reduced by nearly 50 percent. Commanders must develop priorities, accept risk, and require the combat foroe to carry only the bare essentials needed for its own support. Nonessential equipment should be identified, collected, and stored until it is needed. In situations where there are conflicts between the weighl of ammunition and weapons, experience has shown that it is belter lo carry more ammunition and fewer weapons. In the mountains, commanders should strive to achieve the imperatives indicated in Figure 5-2

- Limit supplies to essentials.
- Lighten the individual soldier's combat load.
- Improvise methods and supply sources, to include utilizing captured enemy supplies and equipment.
- Use aviation assets to increase responsiveness.
- Anticipate maintenance requirements.
- Develop plans that place realistic demands on the CSS system.

Figure 5-2. Mountain Supply Imperatives

SECTION II - SUPPLY

5-7. Units operating in mountainous terrain transport supplies by a combination of wheeled vehicles, over snow vehicles, i ndigenous pack animals and personnel (see FM 3-05.27), assisted by Army and Air Force lift assets. These combinations depend on equipment availability, location of combat units, type of terrain, and weather. However, any combination of resupply usually i ndudes combat soldiers man-packing supplies to their positions.

5-8. Since combat operations in the mountains are decentralized, CSS operations are correspondingly decentralized. This decentral ization serves to create heavier man-loads, while rough, steep terrain decreases the amount soldiers are able lo carry. Although most soldiers are eventually able to acdimate themselves to higher elevations, their pace and subsequently the overall pace of the entire operation slows down as elevation increases.

5-9. Mountain warfare is highly dependent on accurate logistical planning if supply operations are to function smoothly. To win in any area of operations (AO), commanders normally seek to move and strike as rapidly as possible. Rapidly changing tadical situations may cause long supply lines, resulting in delay or complete disruption of supply operations. To mitigate these risks, situational understanding, rapid dedsions, and continuous coordination between tactical and logistical planners are essential. Stockpiling and caching supplies may also help to decrease the risks to resuppply.

5-10. The total tonnage of supplies required by the force may also decrease. For example, while individual vehide petroleum, oils, and lubricants (POL) consumption may increase, overall consumption may decrease because of lower vehide movement. The quantity of supplies needed by the individual soldier normally increases. Soldiers consume more food because of increased energy expenditure, and need many additional items of equipment, such as extra clothing, sleeping bags, climbing equipment, tents, and stoves, all of which must be stored and transported.

SUPPLY ROUTES

5-11. Main supply routes are generally limited to the roads located along major valleys and, through necessity, to the smaller, more restrictive trails that follow or parallel the ridgelines. The limited number of routes increases the volume of traffic and places heavy demands on engineer units to maintain them. In most cases, engineer units require assistance in clearing and developing, as well as in securing, these routes. Travel times for ground transportation assets are significantly increased due the generally poor quality of mountain roads and trails, frequent switchbacks, and steep grades that require lower vehicle speeds. Traffic control assumes increased importance due to the limited number of routes in the mountains, and may require an increased number of military police dedicated to the task of battlefield circulation control. In particular-

- Existing roads should be rapidly analyzed for bottlenecks, deployment areas, passing places, and turnarounds for various vehicles.
- Routes should be classified as one- or two-way, and schedules developed for the use of one-way routes.
- Signs should be placed for both day and night moves on difficult and dangerous routes.
- Whenever possible, separate routes should be designated for vehicular and dismounted movement. Additionally, separate routes should be designated for wheeled and tracked vehicles, particularly if the latter are likely to damage road surfaces.

5-12, The enemy will emphasize destroying logistical units and interdicting supply activities. Enemy units will infiltrate and seize key terrain that dominates supply routes in an effort to disrupt and isolate units from their logistics support. Using mountain trails and roads without securing the high ground on both sides invites ambush Patrols must be continually conducted at irregular intervals to verify the status of roads and prevent enemy infiltration. Patrols must be continuously alert for ambush and they must be skilled at locating and identifying mines. However, a combination of patrols and aerial reconnaissance is the best means of providing route security-Observation posts on dominant terrain along supply routes are also essential for early warning of enemy infiltration into rear areas.

5-13. Most often, units have to use the narrrow ridge trails as alternate supply routes, i n some instances as main supply routes, to reduce the volume of traffic on the main supply routes located along valley floors. This involves movements in much more restrictive terrain and exposure to excellent observation and fire by the enemy. Supply columns moving along separate routes face the same problems as combat units; they face the difficulties of being able to provide mutual support due to compartmented terrain, should one column come under attack. Movement of supplies at night may reduce vulnerability to enemy attack, but night marches present other hazards due to the difficult terrain, and require daylight reconnaissance, careful route preparations, and using guides

CLASSES OF SUPPLY

CLASS I: RATI ON S AND WATER

5-14. The strenuous activities required during mountain operations increase caloric requirement to 4,500 calories or more per day. Improper or too little food means soldiers will lack the stamina to accomplish the mission. Although combat rations are normally used, unitized group rations (UGRs) should be provided once a day if the siluation permits. Individual packages of oatmeal and dehydrated soup mixes should be issued if the UGR cycle cannot be maintained.

5-15. In abrupt ascents to high altitude, soldiers do not have time to acclimate themselves, so their entire circulatory system labors to supply oxygen to the body. I n this situation, standard rations are hard to digest and special rations, such as the ration, cold weather (RCW), that allow soldiers to eat light and often should be procured. The totally self-contained operational ration consists of one full day's feeding in a flexible, white-camouflaged meal bag. It contains cooked, freeze-dried, or other low moisture entrees, as well as a variety of items such as oatmeal, a nut-raisin mix, and fruit-cookie bars. The RCW provides sufficient calories (approximately 4.500 kilocalories) to meet the increased energy expenditure during heavy exertion, while limiting sodium and protein content to reduce the risk of dehydration. Because of rapidly changing weather conditions and the difficulty of resupply, each soldier may need to carry two to three days' supply of rations. However, this increases the soldier's load by approximately 10 to 15 pounds.

5-16. Proper water production, resupply, and consumption are essential and a constant challenge during mountain operations. I n low mountains, planners should count on at least four quarts of water per soldier per day when static and up to eight quarts per day when active. In high mountains, planner should increase those requirements by about two quarts per soldier. In the mountain environment, medical care often requires an increased water supply and must be considered as part of the original planning and contingency factors.

5-17. Units should always be prepared to use natural water sources to help reduce the logistics burden However, far above the timberline, water is extremely difficult to find. Special measures must be taken to protect it from freezing in cold weather, such as placing canteens in the chest pockets of the extended cold weather clothing system (ECWCS) coat, hanging a two-quart canteen on a strap under the coat, or utilizing a camel-back type, commercially available, canteen under overgarments. Purification and chemical sterilization are always necessary no matter how clean mountain water may appear. Micro-organisms present in mountain water may cause serious illness and rapidly degrade the strength of a unit. If above ground water sources cannot be located or are not reasonably available, drilling for underground sources may become a critical engineer task. Once engineer units access the water, quartermaster units have responsibility for completing the water points and purifying the water.

CLASS II: GENERAL SUPPLIES

5-18. General supplies indude expendable administrative items, individual dothing and equipment, tentage, and other items authorized by common tables of allowance. All units must deploy with enough Class 11 items lo last until routine resupply can be established. Spedal items, such as extended cold weather dothing, gloves, dimbing equipment, extended cold-weather sleep systems, batteries, and one-burner cook stoves, will be in great demand. Due to the rugged nature of the terrain, mountain operations also increase requirements for replacement items of Individual dothing and equipment. Combat boots, for example, may be expected to last approximately two weeks in harsh rocky terrain.

CLASS III: FUEL AND PACKAGED PETROLEUM PRODUCTS

5-19. Individual vehicles need much more fuel in mountainous terrain. However, limited road nets and steep slopes reduce the volume of vehide traffic and overall fuel consumption. The heavy reliance on aviation assets for resupply and movement increases aviation fuel requirements. A commander must routinely plan for the emplacement of a forward arming and refueling poinl (FARP) within their AO to support intensive aviation operations. Battalions should establish a fuel point in the field trains using collapsible fuel drums. These drums should be operational as soon as the field trains are established and prepared to receive fuel from the forward supply company as soon as it arrives. When terrain makes refueling operations vulnerable to attack, units should conduct forward refueling using supply point distribution, and dispense fuel to unit vehides using the tailgate technique. The lack of suitable terrain normally increases the percentage of forward refueling done by this method

5-20. At 600 meters (2,000 feet), multi-fuel cook staves operate at about 75 percent effidency. When soldiers refuel cook stoves, they must avoid using automotive fuel. Fuel points must supply units with refined or white gasoline that is spedfically produced for pressurized stoves. Relatively large quantities of this fuel will be used when procuring water and preparing food. Adequate quantities of five-gallon cans, nozzles, and one-quart fuel bottles musi be on-hand before deployment.

CLASS IV: CONSTRUCTION, BARRIER, AND FORTIFICATION MATERIALS

5-21. Soldiers should make maximum use of local materials to reduce Class IV requirements and demands on the transportation system. Gabion-type material is espedally versatile during mountain operations. Gabions are widely used in the mountains for constructing obstades, fighting positions, anchors, mountain installations, traverse platforms, and helicopter LZs; for creating landslides or rockfalls; and for repairing roads. Units should stock adequate quantities of easily transportable sizes of reinforcing mesh and other suitable materials for constructing gabions.

CLASS V: AMMUNITION

5-22. Because of terrain, ammunition resupply is difficult, making strict fire control and discipline an absolute necessity. Ammunition transfer poi nts need to be as far forward as possible without revealing friendly unit locations or pfad ng ammunition stocks at risk of capture or destruction. Di rect delivery to the user may be required using aeri al resupply. Innovation and flexibility are critical, in the mountains, the traditional mixes of tank ammunition may be less effedive. Depending upon the spedfic threat, more rounds may be needed to attack light vehicles and fortified positions and less may be needed to engage tanks. Ammunition consumption for direct fire weapons may be low, however, consumption of indirect fire munitions, such as grenades, mortars, and artillery, may be high because of the dead space common to mountainous terrain. Planners must ensure that increased consumption of indirect fire munitions is induded in computating required supply rates.

CLASS VII AND IX: MAJOR END ITEMSAND REPAIR PARTS

5-23. Rugged terrain and dim ate extremes cause an increase in repair parts consumption. However, overall vehide utilization decreases in many situations Because it is difficult to transport large end items to toward units, the commander must place additional emphasis on preventive maintenance and repair.

CLASS VIII: MEDICAL SUPPLIES

5-24. The medical platoon obtains medical supplies for the battalion from the supporting forward support medical company (FSMC) or similar task organized medical element Medical supply organizations may distribute supply by various means: supply point distribution, unit distribution, or a combination of both. Mountainous terrain necessitates using suppty point distribution to a great extent. Medical supply activities must maximize use of empty evacuation assets moving forward to execute unit distribution of supplies as often as possible. The terrain will severely constrain ground movement operations. Pre-planned unit distribution via air assets is a must for emergent situations, such as mass casualty scenarios. Medical supplies must have a high priority for movement. Distribution of Class V! II via air lines of communications (LOCs) should occur as often as tactically feasible.

SECTION III — TRANSPORTATION AND MAINTENANCE

5-25. Transportation assets For mountain operations are often limited, and their use requires sound planning. Although vehicles are used to move supplies as far forward as possible, they may not be able to reach deployed units. Usi ng smai ler cargo vehides with i mproved cross-country mobi lity and dedicated aircraft is paramount to sustaining units in the mountains. Locally obtained animals, indigenous personnel, or combat soldiers must often move supplies from roads and trails to unit positions. The poor quality of road networks requires increased engineer effort. The rugged mountain terrain aids in infiltration increasing security requirementsalong the route.

5-26. Air resupply should always be considered lo reduce the transportation burden on ground assets. Therefore, support personnel should be well-trained in aerial resupply and sling-load operations. Aerial resupply, either by parachute drop, free drop, or cargo helicopter, may be available for a variety of tactical situations. However, unpredictableweather and air currents, cloud cover, and lack of suitable landing zones make aerial delivery unreliable, and higher elevations decrease overall aircraft lift capabilities. The integrated use of available helicopter transport should be used for forward transport of mail, replacements, returnees, and personnel service support providers, such as chaplains and finance support teams.

5-27. Fixing equipment as far forward as possible takes on added importance during mountain operations. In low mountains, equipment recovery and maintenance teams are critical in keeping limited routes clear and returning damaged vehicles to the battle in the shortest possible time. In high mountains, these teams are generally less critical to units operating there because terrain often limits vehicle use. Helicopter repair teams are critical in all mountainous environments due to helicopters flying at or near the maximum limits of their operational capabilities to meet increased needs for helicopter support, 1 n all cases, maintenance turn-around time increases to compensate for fatigue and the other effects of the environment on maintenance personnel. Figure 5-3 lists some of the critical repair parts that often fail or require frequent replacement during mountain operations.

- **Tires**
- **Tracks**
- **NVG Batteries**
- **Communications Equipment**
- **Cooling Systems**

Figure 5-3. Key Repair Parts

5-28. Drivers well trained in proper maintenance and driving techniques eliminate a great deaf of unnecessary maintenance and reduce maintenance requirements and vulnerable bottlenecks. All soldiers must devote increased attention to applicable FMs and TMs for their weapons and equipment and must conduct preventative maintenance, to include the availability and use of suitable cleaning solvents and lubricants, appropriate for the weather and terrain conditions.

SECTION IV - PERSONNEL SUPPORT

5-29. Personnel support in the mountains is not unlike that provided toother lypesof operations except for the limitations on soldiers and equipment posed by the environment. Key personnel support missions are to provide manning and personnel service support to unit commanders, soldiers, and Army civilians.

5-30. Personnel units normally depend on large quantities of automation equipment to successfully accomplish their mission. Adverse weather and rugged terrain may decrease their reliability and commanders should take extra precautions to protect this equipment. Additionally, the difficulty in establishing and maintaining communications may require an increased reliance on manual strength reporting until communications and automated systems are firmly established.

5-31. Postal services establish the link between soldiers and their family and friends and assist in defeating the isolation caused by the compartmented terrain and the resulting dispersion of units. However, the limited lines of communications in mountainous terrain may adversely affect mail distribution. Inefficient distribution of mail can quickly undermine morale, regardless of the AO. The timely delivery or mail is especially important in countering the shock of entering a new environment. Commanders should consider devoting a high priority to the distribution of mail on arrival in the theater of operations. FM 1-0 describes in detail the critical personnel systems and functions essential to providing manning and personnel service support.

SECTION V - COMBAT HEALTH SUPPORT

PLANNING

5-32. Combat health support (CHS) in the mountains is characterized by–

- Difficulty in accessing casualties in rugged terrain.
- Increased need for technical mountaineering skills for casualty evacuation.
- Longer periods of time needed for casualties to be stabilized.

5-33. When planning CHS, commanders must consider the tactical situation, the nature of the terrain, and speed of movement along the chain of evacuation. Aid stations should be centrally located in relation to supported units. The exact location should be based on the ability to provide shelter from the elements, cover and concealment from the enemy, ease of evacuation, and expected casualty rates.

5-34. The decentralization in the mountain area of operations also forces the decentralization of CHS. Commanders may need to establish casualty collection points, operated by aidmen from the evacuation section, to support isolated elements. These points are designated at intermediate points along the routes of evacuation where casualties may be gathered. Additionally, multiple ambulance exchange points may be required to transfer casualties from one type of transportation to another.

EVACUATION

5-35. Aeromedical evacuation remains the preferred form of casualty evacuation in mountain operations. Aircraft provide the best capability of evacuating casualties from isolated locations and transporting them to treatment centers. However, in many instances, even lightly wounded personnel may not be able to move unassisted over rough terrain and LZs may not be available.

5-36, Medical evacuation teams must complete reconnaissance and install necessary evacuation systems along each evacuation route before the onset of casualties. Litter relay stations may be required at predetermined points to conserve the stamina of litter bearers and aocelerate evacuation. The evacuation plan must include measures to care for wounded soldiers at points along the route of evacuation where delays are possible. The plan must also depict all evacuation routes and provide for proper disposition of medical personnel along the lines of evacuation (see FM **4**-02,2). Evacuating the wounded from mountain combat zones normally requires a larger number of medical personnel and litter bearers than on flat terrain. The number and type of evacuation systems depend on mission, enemy, terrain and weather, troops and support available, time available, civil considerations (METT-TC) and the factors listed in Figure **5-4.**

- **The patient's condition**
- **Anticipated casualty rate**
- **Importance of rapid movement**
- **Number of available evacuation teams**
- **Number of evacuation routes available**
- **Transportation assets and equipment available**
- **Availability of suitable and secure LZs**

Figure 5-4. Factors Affecting Evacuation Systems

MOUTAIN EVACUATION TEAMS

5-37. As the battle to control the heights evolves, combatants attempt to exploit technical aspects of terrain. Consequently, commanders must anticipate the need for evacuation teams, normally Level 2 mountaineers, that have the capability to reach, stabilize, and evacuate casualties in rugged terrain. Ground evacuations are generally classified as steep slope (non-technical) or high angle (technical). The mission of trained mountain evacuation teams is to move casualties over cliffs, obstacles, and other technical terrain that would significantly impede the mobility of standard litter bearers. Using evacuation systems to negotiate obstacles shortens routes and i ncreases the speed of evacuation.

5-38. Because units normally deploy over a wide area and the availability of qualified technical evacuation teams is likely to be limited, all soldiers should be trained to conduct less technical, steep-slope evacuations. Two of the most qualified evacuation teams should be identified in each battal ion-si zed unit prior to planned operations. They should be designated as the battalion's technical evacuation assets and should undergo more advanced mountaineering training and rigorous evacuation training. These soldiers can also devetop and teach a program of instruction that will increase the proficiency of the company evacuation teams.

5-39. Mountain evacuation teams must install the necessary evacuation systems before casualties occur. They must man the systems, move casualties over the obstacle until the evacuation route is no longer required, disassemble the system, and redeploy as necessary. Depending on the specific terrain, evacuation teams may demand extensive additional training in some of the areas listed in Figure 5-5.

TREATMENT

5-40. Treatment of Ihe wounded in forward areas by medical personnel is extremely difficult in restrictive terrain, since even a single company is often deployed over a wide area. Combat in the mountains demands a greater reliance on self-aid, buddy-aid, and unit combat lifesaver techniques. Emphasis must be placed on lifesaving and life-preserving measures to be performed before medical personnel arrive. Unit combat lifesavers must be identified and trained to perform in the absence of medics. Units operating in mountainous areas should strive to meet or exceed Army standards for the number of combat lifesavers required for their specific unit See FM 4-02.92 for more information on combat lifesavers.

- High-angle ascending and descending techniques.
- Anchor points and systems.
- Litter rigging.
- Lowering and raising systems.
- Avalanche search and rescue.

Figure 5-5, Mountain Evacuation Team Tasks

5-41. Soldiers in mountain regions are exposed to many and varied types of illnesses and injuries. Appendix A describes the cause, prevention, symptoms, and treatment of common mountain illnesses and injuries.

Appendix A

Mountain Illnesses and Injuries

Table A-1. Chronic Fatigue and Its Effects

CHRONIC FATIGUE (ENERGY DEPLETION)			
CAUSE	**PREVENTION**	**SYMPTOMS**	**TREATMENT**
Low blood sugar. Sources of energy are depleted. Insufficient caloric intake.	Provide adequate food (type and quantities). Monitor food intake and ensure soldiers eat 4,500 calories or more per day. Eat small, frequent meals rather than large, infrequent meals. Snack lightly and often. Increase amounts of fat in diet.	Difficulty sleeping. Fatigue, irritability, and headache. Difficulty thinking and acting coherently -- impaired judgement. Victims begin to stumble and become clumsy and careless. Energy depletion resembles and aggravates hypothermia. The body does not have enough fuel to maintain proper body temperature. As a result inadequate sources of energy, coupled with cold, create a compound or synergistic effect.	Proper diet and rest. Treat synergistic effects if required.

Table A-2. Dehydration and Its Effects

DEHYDRATION			
CAUSE	PREVENTION	SYMPTOMS	TREATMENT
Loss of too much fluid, salt, and minerals due to poor hydration. Contributing Factors: Water loss occurs through sweating, breathing, and urine output. In cold climates, sweat evaporates so rapidly or is absorbed so thoroughly by clothing layers that it is not readily apparent. In cold weather, drinking is inconvenient. Water is hard to resupply, heavy to carry, and freezes in colder climates. Lack of humidity in the dry mountain air. Diminished thirst sensation induced by hypoxia.	Drink 3 to 4 quarts of water per day when static and up to 8 quarts during increased activity. Adequate rest. Avoid caffeine (coffee, tea, soda) and alcohol, as they compound dehydration. Increase command supervision. Keep canteens full. Use flavored powdered drink mixes to encourage water consumption.	Generally tired and weak. Mouth, tongue, and throat become parched and dry, and swallowing becomes difficult. Darkening of urine. Constipation and painful urination. Loss of appetite. Rapid heartbeat. Headache, dizziness, and nausea with or without vomiting. Difficulty focusing eyes. Dehydration compounds the effects of cold and altitude.	Sufficient hydration to offset water loss. Rest. Severe cases may require an IV. Insulate as required and evacuate.

Table A-3. Giardiasis and Its Effects

GIARDIASIS (PARASITICAL ILLNESS)			
CAUSE	PREVENTION	SYMPTOMS	TREATMENT
Parasitical illness contracted from drinking unpurified water.	Drink only potable water. Boil water for 3 to 5 minutes. Use approved water purification tablets or filters. Keep water containers clean.	Abdominal pain. Weakness and nausea. Frequent diarrhea and intestinal gas. Loss of appetite.	Proper hydration with potable water. Evacuation and prescribed medications.

Table A-4. Hypoxia and Its Effects

HYPOXIA			
CAUSE	PREVENTION	SYMPTOMS	TREATMENT
Rapid ascent to high altitudes (above 3,000 to 4,000 meters or 10,000 to 13,000 feet).	Acclimatization. Slow ascent. Limited activities. Lonq rest periods.	Impaired judgment, perception, and higher mental functions increasing with altitude.	Evacuation to tower altitude.

Table A-5. Acute Mountain Sickness (AMS) and Its Effects

ACUTE MOUNTAIN SICKNESS (AMS)			
CAUSE	PREVENTION	SYMPTOMS	TREATMENT
Rapid ascent to high altitudes {2,400 meters or 8.000	Acclimatization. Staged and/or graded ascent. During stops, no strenuous activity and only mild activity with frequent rest periods. Increased carbohydrate intake (whole grains, vegetables. peas and beans, potatoes, fruits. honey, and refined sugar). Acetazol amide prescribed by a physician.	Headache and fatigue. Insomnia. irritability, and depression. Coughing and shortness of breath. Loss of appetite, nausea, and vomiting. Dizziness. Swelling ol the eyes and face.	Stop and rest Symptoms will normally subside in 3-7 days if soldiers do not continue to ascend. Observe tor the development of HAPE or HACE. If symptoms do not disappear, a rapid descent of 150 to 300 meters (500 to 1,000 feet) or greater is necessary. Re-ascent should take place only after symptoms are resolved.

Table A-6. High Altitude Pulmonary Edema (HAPE) and Its Effects

HIGH ALTITUDE PULMONARY EDEMA (HAPE)			
CAUSE	**PREVENTION**	**SYMPTOMS**	**TREATMENT**
Unacclimatized soldiers rapidly ascending to high altitudes (2,400 meters or 8,000 feet)*. Acclimatized soldiers ascending rapidly from a high to a higher altitude. Usually begins within the first 2 - 4 days after rapid ascent and generally appears during the second night of sleep at high or higher altitudes. Fluid accumulation in the lungs.	Acclimatization. Staged and/or graded ascent. Sleeping at the lowest altitude possible. Slow assumption of physical activity. Protection from the cold.	Wheezing and coughing (possibly with pink sputum). Gurgling sound in chest. Difficulty breathing. Coma Death may occur if rapid descent is not initiated.	Rapid evacuation recommended. Observe for the development of HACE. Seek qualified medical assistance.
*HAPE most often [illegible] 3 -300 [illegible] 000 [illegible]			

Table A-7. High Altitude Cerebral Edema (HACE) and Its Effects

HIGH ALTITUDE CEREBRAL EDEMA (HACE)			
CAUSE	**PREVENTION**	**SYMPTOMS**	**TREATMENT**
Unacclimatized soldiers rapidly ascending to high altitudes (2,400 meters or 8,000 feet)*. Acclimatized soldiers ascending rapidly from a high to a higher altitude. Excessive accumulation of fluid in the brain.	Acclimatization Staged and/or graded ascent. Slow assumption of physical activity. Protection from the cold.	Most severe high altitude illness. Severe headache, nausea, and vomiting. Staggering walk/sway. Confusion, disorientation, and drowsiness. Coma, usually followed by death.	Immediate evacuation; preferably by air evacuation. Seek qualified medical assistance.
*HACE, like HAPE, most often does not occur until above [illegible],000 meters (12,000 feet)			

Appendix B

Forecasting Weather in the Mountains

The Air Force provides the bulk of the weather support required by the Army; however, reports from other branches of the military service, our own National Weather Bureau, or a foreign country's weather service can also aid in developing accurate forecasts {see FM 2-33.2). Weather at different elevations and areas, even within the same general region, may differ significantly due to variations in cloud height, temperature, winds, and barometric pressure. Therefore, general reports and forecasts must be used in conjunction with the locally observed weather conditions to produce reliable weather forecasts for a particular mountain area of operations.

INDICATORS OF CHANGING WEATHER

MEASURABLE INDICATORS

B-1. In the mountains, a portable aneroid barometer, thermometer, wind meter, and hygrometer are useful to obtain measurements that will assist in forecasting the weather. Marked or abnormal changes within a 12-hour period in the indicators listed in Figure B-1 may suggest a potential change in the weather.

- Barometric Pressure
- Wind Velocity
- Wind Direction
- Temperature
- Moisture Content of the Air

Figure B-1. Measurable Weather Indicators

CLOUDS

B-2 Clouds are good indicators of approaching weather conditions. By reading doud shapes and patterns, observers can forecast weather even without additional equipment.

B-3. Shape and height are used to identify douds. Shape provides information about the stability of the atmosphere, and height above ground level provides an indication of the distance of an approaching storm. Taken together, both indicate the likelihood of predpitation (see Figure B-2). The heights shown in the figure are an estimate and may vary, based on geographical location.

Clouds by Shape

B-4. Clouds may be classified by shape as cumulus or stratus.

- Cumulus douds are often called "puffy" clouds, looking like tufts of cotton. Their thickness (bottom to top) is usually equal to or greater than their width. Cumulus douds are primarily composed of water droplets that cause them to have sharp, distinct edges. These clouds usually indicate instability at the altitude of the atmosphere where they are found. The stormy weather associated with cumuius clouds is usually violent with heavy rains or snow and strong, gusty winds. Precipitating cumulus douds are called cumulonimbus.

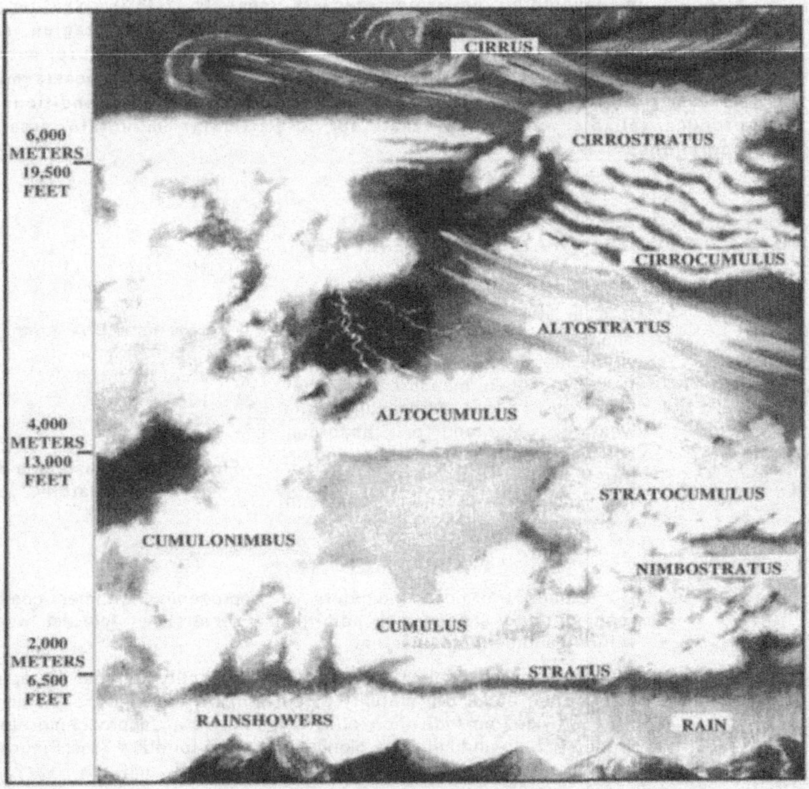

Figure B-2. Types of Clouds

- Stratus clouds are layered, often appearing flattened, with greater horizontal than vertical dimensions. They usually indicate a stable atmosphere, but can indicate the approach of a storm. Stormy weather assodated with stratus clouds usually does not normally include violent winds, and precipitation is usually light but steady, lasting up to 36 hours. Lightning is rarely associated with stratus douds, however, slee! may occur. Fog is also associated with the appearance of stratus clouds. Precipitating stratus clouds are called nimbostratus. and clouds that cannot be determined as stratus or cumulus are referred to as stratocumulus. These latter types may be evolving from one type to another, indicating a change in atmospheric stability.

Clouds by Height

B-5. Clouds are also classified by the height of their base above ground level into three categories- low. middle, and high.

- Low clouds, below 2,000 meters (6,500 feet), are either cumulus or stratus, or their predpitating counterparts. Low douds may be identified by their height above nearby surrounding relief of known elevation. Most precipitation originates from low douds because rain and snow from higher clouds usually evaporates before reaching the ground. As such, low douds usually indicate predpitation. especially if they are more than 1.000 meters (3.000 feet) thick (douds that appear dark at the base usually are at least that thick).
- Middledouds, between 2.000 and 6,000 meters (6,500 and 19,500 feet) above ground, have a prefix of "alto", and are called either altostratus or altocumulus. Middle douds appear less distinct than low douds because of their height. Warm "alto" douds have sharper edges and are composed mainly of water droplets. Colder douds, composed mainly of ice crystals, have distinct edges that grade gradually into the surrounding sky. Middle clouds indicate potential storms, though usually hours away. Altocumulus clouds that are scattered in a blue sky are called "fair weather" cumulus and suggest the arrival of high pressure and clear skies. Lowering altostratus douds with winds from the south indicate warm front conditions, decreasing air pressure, and an approaching storm system within 12 to 24 hours.
- High douds, higher than 6,000 meters (19,500 feet), are drrus. drrostratus, and drrocumulus. They are usually frozen douds with a fibrous strudure and blurred outlines. The sky is often covered with a thin veil of drrus that partly obscures the sun or, at night, produces a ring of light around the moon. The arrival of drrus indicates moisture aloft and the approach of a storm system. Predpitation is often 24 to 36 hours away. As the storm approaches, the drrus thickens and lowers becoming altostratus and eventually stratus. Temperatures warm, humidity rises, and winds approach from the south or southeast.

Other Clouds

B-6 Somedoudsindicate serious weather ahead.

- Towering cumulus douds have bases below 2,000 meters (6,500 feet) and tops often over 6,000 meters (19.500 feet). They are the most dangerous of all types and usually do not occur when temperatures at the surface are below 32-degrees Fahrenheit. They indicate extreme instability in the atmosphere, with rapidly rising air currents caused by solar heating of the surface or air rising over a mountain barrier. Mature towering cumulus douds often exhibit frozen stratusdouds at their tops, producing an "anvil head" appearance. Towering cumulus clouds may be local in nature, or they may be associated with the cold front of an approaching storm. The iatter appears as an approaching line of thunderstorms or towering cumulus clouds. Towering cumulus clouds usually produce high, gusty winds, lightning, heavy showers, and occasionally hail and tornadoes (although tornadoes are rare in mountainous terrain). Such thunderstorms are usually short-lived and bring clear weather.

G Cloud caps often form above pinnacle and peaks, and usually indicate higher winds aloft. Cloud caps with a lens shape (similar to a "flying saucer") are called lenticular and indicate very high winds (over 40 knots). Cloud caps should always be watched for changes. If they grow and descend, bad weather can be expected.

APPLYING THE INDICATORS

B-7. Weather forecasts are simply educated estimations or deductions based on general scientific weather principles and meteorological evidence. Forecasts based on past results may or may not be accurate. However, even limited experience in a particular mountainous region and season may provide local indications of impending weather patterns and increased accuracy. Native weather lore, although sometimes greatly colored and surrounded in mystique, should not be discounted when developing forecasts, as it is normally based on the local inhabitants' long-term experience in the region.

BAD WEATHER

B-8. Signs of approaching bad weather (within 24 to48 hours) may include—

- A gradual lowering of the clouds. This may be the arrival or formation of new lower strata of clouds. It can also indicate the formation of a thunderhead.
- An increasinghaloaroundthesunorlhe moon.
- An increase in humidity and temperature.
- Cirrus clouds.
- A decrease in barometric pressure (registered as a gain in elevation on an altimeter).

STORM SYSTEMS

B-9. The approach of a storm system is indicated when—

- A thin veil of cirrus clouds spreads over the sky, thickening and lowering until altostratus clouds are formed. The same trend is shown at night when a halo forms around the moon and then darkens until only the glow of the moon is visible. When there is no moon, cirrus clouds only dim the stars, but altostratus clouds completely hide them.
- Low clouds, which have been persistent on lower slopes, begin to rise at the time upper clouds appear.

- Various layers of clouds move in at different heights and become abundant.
- Lenticular clouds accompanying strong winds lose their streamlined shape, and other doud types appear in increasing amounts,
- A change in the direction of the wind is accompanied by a rapid rise in temperature not caused by solar radiation. This may also indicate a warm, damp period.
- A light green haze is observed shortly after sunrise in mountain regions above the timber line.

THUNDERSTORMS

B-10. Indications of local thunderstorms or squally weather are—

- An increase in size and rapid thickening of scattered cumulus clouds during the afternoon.
- The approach of a line of large cumulus or cumulonimbus clouds with an "advance guard" of altocumulus douds. At night, increasing lightning windward of the prevailing wind gives the same warning.
- Massive cumulus douds hanging over a ridge or summit (day or night).

STRONG WINDS

B-11. Indications of approaching strong winds may be—

- Plumes of blowing snow from the crests of ridges and peaks or ragged shreds of doud moving rapidly.
- Persistent lenticular douds, a band of douds over high peaks and ridges, or downwind from them.
- A turbulent and ragged banner doud that hangs to the lee of a peak.

PRECIPITATION

B-12. When there is predpitation and the sky cannot be seen—

- Small snowflakes or ice crystals indicate that the douds above are thin, and fair weather exists at high elevations.
- A steady fall of snowflakes or raindrops indicates that the predpitation has begun at high levels, and bad weather is likely to be encountered on ridges and peaks.

FAIR WEATHER

B-13. Continued fair weather may be assodated with—

- A doudless sky and shallow fog, or layers of haze at valley bottoms in early morning.
- A doudless sky that is blue down to the horizon or down to where a haze layer forms a secondary horizon.
- Conditions under which small cumulus douds appearing before noon do not increase, but instead decrease or vanish during the day.

- Clear skies except for a law cloud deck that does not rise or t hick en during the day.

B-14. Signs of approaching fair weather include—

- A gradual rising and diminishing of clouds.
- A decreasing halo around the sun or moon.
- Dew on theground in the morning.
- Small snowfiakes, ice crystals, or drizzle, which indicate that the clouds are thin and fair weather may exist at higher elevations.
- An increase in barometric pressure (registered as a loss in elevation on an altimeter).

GLOSSARY

ABN	airborne
acclimatization	the physiological changes that allow the body to adapt or get used to the effects of a new environment, especially low oxygen saturation at higher elevations
ACE	armored combat earth mover
acetazolamide	a pharmaceutical drug used to accelerate acclimatization
ADA	air defense artillery
ADAM	area denial artillery munitions
AH-64	attack helicopter also called the Apacbe
aid	in mountaineering, a climbing device, such as pitons, bolts, chocks, and stirrups, used for body support and upward progress; also used for artificial height in the absence of handholds and footholds
ALOC	air lines of communications
AM	amplitude modulation
ambient temperature	encompassing atmosphere
AMS	acute mountain sickness
anchor	a secure point (natural or artificial} to which a person or rope can be safely attached
aneroid	using no liquid
ANZAC	Australia and New Zealand Corps
AO	area of operations
apnea	temporary suspension of respiration
ARSOF	Army special operations forces
ART	Army tactical task
ARTEP	Army training and evaluation program
assault climber	military mountaineer possessing advanced (Level 2) skills, capable of leading small teams over class 4 and 5 terrain and supervising rigging/operation of all basic rope systems
AT4	a man-portable, lightweight, self-contained, antiarmor weapon
ATGM	anti-tank guided missile
basic mountaineer	a military mountaineer trained in fundamental (Level 1) travel/climbing skills necessary to move safely and efficiently in mountainous terrain

belay	a rope management technique used to ensure that a fall taken by a climber can be quickly arrested: belay techniques are also used for additional safetyfcontrol in rappelling. raising and lowering systems, and for mountain stream crossings
BFV	Bradley fighting vehicle
BSFV	Bradley Stinger fighting vehicle
BN	battalion
&.	command and control
CAFAD	combined arms for air defense
CFV	cavalry fighting vehicle
CHS	combat health support
CNR	combat net radio
continental climate	bitterly cold winters, extremely hot summers; annual rain and snowfall is minimal and often quite scarce for long periods
Cordillera	principal mountain ranges of the world, named after the Spanish word for rope
crampons	climbing irons, attached to the bottom of boots, used on ice or snow in mountaineering
crevice	a narrow opening resulting from a split or crack as in a cliff
CS	combat support
CSS	combat service support
DA	Department of the Army
defile	a narrow passage or gorge
DPI CM	dual-purpose improved conventional munition
DZ	drop zone
ECWCS	extended cold weather clothing system
edema	a local or general condition in which the body tissues contain an excessive amount of tissue fluid
evacuation team	a team trained to move casualties over steep slopes, cliffs, and other obstacles that would significantly impede the mobility of standard litter bearers
EW	electronic warfare
F	Fahrenheit
FARP	forward arming and refueling point
FASCAM	family of scatterable mines
FEBA	forward edge of the battle area

fixed alpine path	a mountain path created by any combination of aids, to include steps, stanchions, standoff ladders, suspended walkways, cableways, or other improvements made of materials available; normally an engineering task.
fixed rope	a rope, or series of ropes, anchored to the mountain at one or more points to aid soldiers over steep, exposed terrain; usually installed by lead climbing teams (normally Level 2 qualification)
flash defilade	to arrange fortifications to protect from fire
FM	field manual; frequency modulation
FSMC	forward support medical company
FOX system	a lightly-armored, wheeled laboratory that takes air, water, and ground samples and immediately analyzes them for signs of weapons of mass destruction
gabion	a wicker basket filled with earth and stones often use in building fortifications; can also be created out of similar materials, such as wire mesh/fence, lumber, plywood, or any suitable material that forms a stackable container for rocks, gravel, and soil
giardiasis	parasitical illness
glaciated	covered with glacial ice
GPS	global positioning system
GTA	graphictraining aid
guide	a soldier experienced in all aspects of mountaineering who has the skills and knowledge to identify obstacles and ways to overoome them; commander's advisor on technical mountaineering matters that could affect the tactical scheme of maneuver; primary function of mountain leaders (Level 3 qualification)
HACE	high altitude cerebral edema
HAPE	high altitude pulmonary edema
HE	high explosives
Hellfire	tank-killing missile carried by the Apache attack helicopter
high mountains	mountains that have a local relief usually exceeding 900 meters (3,000 feet)
HUMINT	human intelligence
HWY	highway
hygrometer	an instrument used to measure humidity or moisture content in the air
hypoxia	a deficiency of oxygen reaching the tissues of the body
ice fog trails	steam/smoke trails created by firing weapons
ID	infantry division

IFV	infantry fighting vehicle
IHFR	improved high frequency radio
IMINT	imagery intelligence
installation team	a team organized to construct and maintain rope installations used to facilitate unit movement; usually comprised of Level 1 and 2 mountaineers
interdiction	to stop or hamper
ionospheric	a part of the earth's atmosphere of which ionization of atmospheric gases affects the propagation of radio waves; starts at about 30 miles above ground
IPB	intelligence preparation of the battlefield
IV	intravenous
JSTARS	joint surveillance, target attack radar system
km	kilometer
lead climbing team	a roped dimbing team (usually Level 2 qualification) trained to lead on dass 4 and 5 terrain; establishes/prepares the entire route for the remainder of the unit
leeward	the side sheltered from the wind
lenticutar	having the shape of a double-convex lens
LOC	line of communication
local relief	the difference in elevation between valley floors and the surrounding summits
look-down angles	the angle from the aircraft to the target
low mountains	mountains that have a local relief of 300 to 900 meters (1,000 ta 3,000 feet)
LPT	logistics preparation of the theater
LRS	long-range surveillance
LRSU	long-range surveillance unit
LSDIS	light and spedal division interim sensor
LTC	lieutenant colonel
I 7	landing zone
MANPADS	man-portable air defense system
maritime climate	mild temperatures with large amountsof rain or snow
MBA	main battlearea
METT-TC	mission, enemy, terrain and weather, troops and support available, time available, dvil considerations
MK-19	40-mm grenade machine gun, MOD 3

MOPP	mission-oriented protective posture
motti	Finnish word meaning "a pile of logs ready to be sawed into lumber"; used in military terms to descri be setting the condit ions so a larger force can be defeated in detai l
mountain leader	a military mountaineer possessing the highest level (Level 3) of mountaineering skills with extensive experience in a variety of mountain environments in both winter and summer months
WISE	mobile subscriber equipment
MSRT	mobile subscriber radio terminal
MTF	manual terrain following
NBC	nuclear, biological, and chemical
OCOKA	observation and fields of fire, cover and concealment, obstacles, key terrain, and avenues of approach
OH-58D	a scout and attack helicopter known as the Kiowa Warrior
OP	observation post
OPORD	operation order
OPSEC	operations security
OR	operational readiness
orographic	pertaining to the physical geography of mountains and mountain ranges
PADS	Position Azimuth Determining System
POL	petroleum, oils, and lubricants
protection	in mountaineering, special anchor points established during a roped party dimb to limit potential fall distances, protecting dimbers from severe fall/ground-fall consequences
PSYOP	psychological operations
RAAMS	remote antiarmor mine system
rappel	method of controlled fridional descent down a rope
RCW	ration, cold weather
rockfai l	rockfall occurs on alt steep slopes. It is caused by other dimbers or by the continual erosion of the rock on a mountainside resulting from freezing, thawing, and heavy rain; grazing animals; or enemy action.
SATCOM	satellite communications
scree	small unconsolidated rocks or gravel, fist-**si2e** or smaller, located mostly below rock ridges and diffs
screen i ng crest	a hill or ridge located in front of a radar set to mask it from unwanted returns (dutter) at dose range, and to provide security

	against electronic detection or jamming; screening crest also prevents visual observation and attack by direct fires
SEE	small emplacement excavator
SHELREP	shelling report
SH ORAD	short-range air defense
SINCGARS	Single-channel Ground and Airborne Radio System
SOP	special operations forces
sq u a 11	a sudden, violent wind
SR	special reconnaissance
talus	accumulated rock debris that is much larger than scree, usually basket ball-size or larger
TBP	to be published
TC	[raining circular
TCF	tactical combat force
TCP	trafficcontrd point
temperature inversion	when the temperature is warmer at higher elevations than lower elevations
TM	technical manual
TOC	tactical operations center
TOW	tube-launched, optically tracked, wire-guided, heavy antitank missile system
TRADOC	United States Army Training and Doctrine Command
tundra	treeless, black, mucky soil with permanently frozen subsoil; located in mountainous regions above the timberline
tussocks	grassy clumps
UAV	unmanned aerial vehicle
UGR	unitized group ration
UHF	ultrahigh freciuoncy
Venturi effect	as a fluid (such as air) flows through a constriction (like a mountain pass), the speed increases and the pressure drops
VFR	visual flight rules
VT	variabletime
wind chi11	the rate at which a man or object cools to the ambient temperature; wind increases the rate of cooling and adds to the risk of frostbite, hypothermia, and other cold-weather injuries
wi ndward	beingin or fadngthedirection from which thewind isblowing
WP	white phosphorous

Bibliography

The bibliography lists field manuals by new number followed by old number.

ARMY PUBLICATIONS

Most Army doctrinal publications are available online:
http://155.217.58.58/atdls.htm

AR 385-10. *The Army Safety Program.* 23 May 1988.

FM 1-0 (12-6). *Personnel Doctrine.* 09 September 1994.

FM 2-01.3 (34-130). *Intelligence Preparation of the Battlefield.* 08 July 1994.

FM 2-33.2 (34-81). *Weather Support for Army Tactical Operations.* 31 August 1989.

FM 2-33.201 (34-81-1). *Battlefield Weather Effects.* 23 December 1992.

FM 3-0 (100-5). *Operations.* 14 June 1993.

FM 3-01.8 (44-8). *Combined Arms for Air Defense.* 01 June 1999.

FM 3-01.43 (44-43). *Bradley Stinger Fighting Vehicle Platoon and Squad Operations.* 03 October 1995.

FM 3-01.44 (44-44). *Avenger Platoon, Section, and Squad Operations.* 03 October 1995.

FM 3-04.100 (1-100). *Army Aviation Operations.* 21 February 1997.

FM 3-04.203 (1-203). *Fundamentals of Flight.* 03 October 1988.

FM 3-05.27 (31-27). *Pack Animals in Support of Army Special Operations Forces.* 15 February 2000.

FM 3-09.4 (6-20-40). *Tactics, Techniques, and Procedures for Fire Support for Brigade Operations (Heavy).* 05 January 1990.

FM 3-09.12 (6-121). *Tactics, Techniques, and Procedures for Field Artillery Target Acquisition.* 25 September 1990.

FM 3-09.40 (6-40). *Tactics, Techniques, and Procedures for Field Artillery Manual Cannon Gunnery.* 23 April 1996.

FM 3-09.42 (6-20-50). *Tactics, Techniques, and Procedures for Fire Support for Brigade Operations (Light).* 05 January 1990.

FM 3-11.4 (3-4). *NBC Protection.* 29 May 1992.

FM 3-11.5 (3-5). *NBC Decontamination.* 28 July 2000.

FM 3-]] ,6 (3-6), *Field Behavior of NBC Agents (Including Snufke and Incendiaries).* 03 November (986.

FM 3-11.19 (3-19). *NBC Reconnaissance.* 19 November 1993.

F M 3-21.7 (7-7). ***Tile Mechanized Infantry Platoon and Squad <APC).*** 15 ***Match*** 1985.

FM 3-21,20(7-20), *The Infantry Battalion.* 06 April 1992,

FM 3-21.30 (7-30), *The Infantry Brigade.* 03 October 1995.

F M 3-21.38 (S7-38). ***Pathfinder Operations.*** 09 April 1993.

FM 3-23.10 (23-10). *SniperTraining.* 17 August 1994.

FM 3-23,30 (23-30), *Grenades and Pyrotechnic Signal).* 27 Decern be r 1988,

FM 3-24.3 (20-3). *Camouflage. Concealment, and Decoys.* 30 August 1999.

FM 3-24.32 (20-32). ***Mine/Countermine Operations.*** 29 May 1998.

FM 3-25,9(23-9), *MI6A) and M16A2 Rifle Marksmanship.* 03 July 1989.

FM 3-25.18(21-18). *Foot Marches.* 0 J June 1990.

FM 3-25.26 (21-26). *Map Reading and Land Navigation.* 07 May 1993.

FM 3-25,60(21-60), *Visual Signals.* 30 September 1987,

FM 3-25,76 (21-76). *Survival.* 05 June 1992.

FM 3-34.1 (90-7). ***Combined Arms Obstacle Integration.*** 29 September 1994.

FM i 34.2 (9(1 13 I 1. ***Combined Arms Brea, lung Operations.*** -N iilmi.iri I V) I .

FM 3-34.102 (5-102), *Cotmtermobility.* I4March 1985.

FM 3-34.112(5-103). *Survivability.* 10 June 1985.

FM 3-34.214 (5-250). *Explosives and Demolitions.* 30 July 1998.

FM 3-34,223 (5-7-30). *Brigade Engineer and Engineer Company Combat Operations I Airborne, Air Assault, Light}.* 28 December 1994.

FM 3-34.330 (5-33). *Terrain Analysis.* I 1 July 1990.

FM 3-60 (6-20-10). ***Tactics. Techniques, and Procedures for the Targeting Process.*** 08 May 1996.

FM 3-90.21 (6-20-1). ***Tactics, Techniques, and Procedures for the Field Artillery Cannon Battalion.*** 29 November 1990.

FM 3-91 -2 (7 I -2). ***The Tank and Mechanized Infantry Battalion Task Force.*** **21** September 1988.

FM 3-91,3 (71-3), *The Armored and Mechanized Infantry Brigade.* 08 January 1996,

FM 3-9], 323(71-123). *Tactics and Techniques for Combined Arms Heavy Forces: Armored Brigade, Battalion Task Force, and Company Team,* 30 September 1992,

FM 3-97.3 (90-3). *Desert Operations. 24* Augusl 1993.

FM 3-97.1 I (90-1 I). *Cold Weather Operations. TUP.*

FM 3-97,22 (90-22), *(Night) Multi-Service Night and Adverse Weather Combat Operations.* 31 January 1991.

FM 3-97,50 (3-50), *Smoke Operations,* 04 December 1990,

FM 3-100.14 (100-14). *Risk Management,* 23 April 1998.

FM 3-100.15 (100-15). *Corps Operations.* 29 Oelnber 1996.

FM 3-100,40(100-40), *Tactics,*TBP,

FM 3-100.55 (100-55). *Reconnaissance and Surveillance.* TBP,

FM 3-100.71 (71-100). *Division Operations.* 28 August 1996.

FM 4-02.2 (8-10-6). *Medictd Evacuation in a Theater of Operations Tactics. Techniques, and Procedures.* 14 April 2000.

FM 4-02,22 (22-51), *Leaders' Manual for Combat Stress Control.* 29 September 1994.

FM 4-02.92 (8-10-4). *Medical Planum Leaders' Handbook Tactics, Techniques, and Procedures.* I6November 1990,

FM 4-20.1 (10-27). *General Supply in Theaters of Operations.* 20 April 1993.

FM 4-20.2 (10-23). *Basic Dtfcrrine for Artnv Field Feeding and Class I Operations Management.* IK April 1996.

FM 4-25,10 (21-10), *Field Hygiene and Sanitation.* 21 June 2000

FM 4-25.11(21-11). *First Aid for Soldiers.* 27 October 1988,

FM 4-30.32 (9-207). *Operations and Maintenance of Ordnance Materiel in Cold Weather.* 20 March 1998.

FM 5-0 (101-5). *Sniff Organization and Operations.* 31 May 1997.

FM 6-0 (100-34). *Command and Control.* TBP.

FM 6-02.1 I (24-1 I), *Tactical Satellite Communications.* 20 September 1990.

FM 6-02.18 (24-1 8). *Tactical Single-Channel Radio Communications Techniques.* 30 September 1987.

F'M 6-02.32 (I 1-32). *Combat Net Radio Operations.* 15 Otlciber 1990.

FM 6-02,43(1 1-43), *The Signal Leader's Guide,* 12 June 1995,

FM 6-02,55 (11-55). *Mobile Subscriber Equipment < MSB) Operations. 22* June 1999.

FM 6-22 (22-100), *Army Leadership Be, Know, Do.* 31 August 1999,

FM 7-0 (25-100). *Training die Farce.* 15 November 1988.

FM 7-10(25-101). *Battle Focused Truining.* 30 September 1990.

FM 7-I5(25-XX). *Army Universal Task List.* TBP,

FM 21-305. *Manna! far the Wheeled Vehicle Driver.* 27 August 1993.

GTA 8-5-60. *A Soldier's (hade to Staying Healthy at High Elevations.* 02 September 1996.

GTA 8-6-12, *Adverse Effects of Cold.* 01 August 1985.

TC3-10. *Commander's Tactical NBC Handbook.* 29 September 1994.

TC 24-20. *Tactical Wire and Cable Techniques.* 03 October 1 988.

TC 24-21. *Tactical Multichannel Radio Communications Techniques.* 03 October 1988.

TC 90-6-1. *Military Mountaineering.* 26 April 1989.

NONMILITARY PUBLICATIONS

Jalali. Ali Ahmad and Lester W. Grau. *The Other Side of the Mountain.* Ouanlko, Va.: Marine Corps Combat Development Command, USMC Studies and Analysis Division, 1999.

Ellis, Robert B. *See Naples and Die.* London: McFarland & Co. 1996.

Gawryeh, George W. "The Rock of Gallipoli" in *Studies in Battle Command.* Kurt Leavenworth. Kans.: LIS Army Command and General Staff College, Combat Studies Institute. 1995.

Cirau, LesLcr W. *The Bear Went Over the Mountain: Soviet Combat Tactics in Afghanistan.* Washington. D.C: National Defense University Press, 1996.

DOCUMENTS NEEDED

These documents should be available to the intended users of this manual.

DA Form 2028, Recommended Changes to Publications and Blank Forms. I February 1974,

FM 7-0 (25-100). *Training die Force.* 15 November 1988.

FM 7-10 (25-101). *Battle Focused Training.* 3(1 September 1990.

FM 3-0 (100-5), *Operations.* 14 June 1993,

FM 3-40(100-40). *Tactics.* TBP.

TC 90-6-1. *Military Mountaineering.* 26 April 1989.

Index

Entries are by paragraph number unless stated otherwise.

Made in United States
North Haven, CT
04 November 2026

21774392R00143